IEE Electrical measurement series 2
Series Editors: A. C. Lynch
A. E. Bailey

MICROWAVE IMPEDANCE MEASUREMENT

Previous volumes in this series

Volume 1 Lock in amplifiers: principles and applications
M. L. Meade

MICROWAVE IMPEDANCE MEASUREMENT

P.I. Somlo
J.D. Hunter

Peter Peregrinus Ltd on behalf of the Institution of Electrical Engineers

Published by: Peter Peregrinus Ltd., London, UK.

British Library Cataloguing in Publication Data

Somlo, P.I.
Microwave impedance measurements. — (IEE electrical measurement series; 2)
1. Microwave measurements 2. Impedance (Electricity) — Measurement
I. Title II. Hunter, J.D. III. Series
621.381'37 TK7876

ISBN 0-86341-033-2

Printed in England by Short Run Press Ltd., Exeter

Contents

Preface

This book is a survey of microwave impedance measuring techniques, and is intended to be both a guide to the metrology student, and a framework against which the practising metrologist can evaluate new developments.

It is assumed that the reader has a basic understanding of the theory of electromagnetic waves and transmission lines, and also has some algebraic ability, although the mathematics used has intentionally been kept as simple as possible. The subject matter covers fundamental principles, network analysis techniques, measurement methods and circuit descriptions. Measurement methods based on a common principle have been grouped in chapters. Because any metrologist soon modifies a measurement procedure to suit his particular task and equipment, a vast literature on impedance measurements exists in several languages, and it is not possible to discuss every implementation of a particular measurement method. The references cited in this book are intended to be a representative sample of this literature.

The measurement methods discussed range from the earliest standing wave experiments of Lecher in the 19th-century to the comparatively embryonic implementation of six-port theory, although the detailed description of automatic measuring systems is outside the scope of this book. Recently adopted analytical techniques, such as matrix renormalization are described, as are both real and synthetic pulse time-domain measurements.

We thank the management of the Commonwealth Scientific and Industrial Research Organization Division of Applied Physics for permission to write this book, and the typing and drawing office staff for their support in its preparation. We acknowledge the debt we owe to the accumulated experience shared so willingly by numerous colleagues at the CSIRO National Measurement Laboratory, and in particular D. L. Hollway (now retired). In so far as some of the measurement techniques described in this book were developed at the National Measurement Laboratory, we acknowledge the skill and dedication with which P. W. Campbell produced many special microwave components.

P. I. SOMLO and J. D. HUNTER

Chapter 1

Basic concepts

1.1 Introduction

Electromagnetic circuits may be characterized as lumped, distributed or a combination of both. The concept of a lumped circuit, in which the elements can be regarded as acting at a point, is particularly useful when the signal wavelength is very much greater than the size of circuit elements and their connecting circuitry. When this condition is not met, the action of circuit components on the propagating signal must be considered to be distributed throughout the physical extent of the components.

The term 'microwave' is a loose description of a frequency range which for our purposes shall extend from a few hundred megahertz to several tens of gigahertz. Because the wavelengths at microwave frequencies are usually comparable with, or much smaller than, the physical size of circuit components, the concept of distributed circuits is essential in a circuit analysis. In particular the transmission lines connecting circuit elements are themselves distributed circuits.

The term 'impedance' has an evolutionary history of its own. Its current definition is a complex ratio of a force-like quantity (force, pressure, voltage, temperature or electric field strength) to a related velocity-like quantity (velocity, volume velocity, current, heat flow or magnetic field strength) [1].

The word impedance was used in 1886 by Sir Oliver Lodge to define what is now called the magnitude, or absolute value of impedance. Despite the fact that Steinmetz [2] introduced the complex definition of impedance before the turn of the 20th-century, in books as late as 1933 by Pohl [3] and 1941 by Hutchinson [4], impedance was still defined to have the meaning given by Lodge.

Impedance is, in general, a function of frequency and the mode of propagation. It can also be a function of direction. If in an electromagnetic field both the electric and magnetic field vectors are given in terms of their co-ordinate components, impedances may be defined in all directions of the co-ordinate axes [5]. However, the impedance in a particular direction cannot be deduced from the vector addition of impedances in other directions. In this book, impedance is understood to be that applying in steady-state conditions. The impedance of a transmission line is that associated with the mode of interest when looking along the transmission line in the direction of interest.

Nowadays impedance is often used to refer not only to a property of a device, but to the device itself. Although it is more correct to refer to a device having impedance as an impedor (c.f. resistance–resistor, inductance–inductor, capacitance–capacitor), the dual function of the word is widely accepted and rarely leads to confusion.

For reasons of convenience, the inverse of impedance, called admittance, is often used. In order to refer to both quantities in a unified manner, Bode [6] coined two combinations of these words and suggested 'adpedance' or 'immittance', of which the latter has found common acceptance over the years.

The importance of impedance measurement is best illustrated by examples.

(a) Transmission of power. As is well known, maximum power transfer from a source to a load takes place if the impedances of these are complex conjugates.
(b) Signal reflection. If the impedances of both a source and load are the same as the characteristic impedance of the intervening transmission line, there are no reflected signals on the line, and the source and load are said to be matched. If both source and load are mismatched to the transmission line, multiple reflections will occur, and result in undesirable echoes in voice transmission, ghosts on television pictures, or false pulses in digital transmissions.
(c) Network characterization. Some network properties, such as attenuation, can be defined only when the port impedances of the embedding network are specified [7].
(d) Measurement of material properties. Some material properties, such as dielectric constant and moisture content, can be conveniently determined from impedance measurements.

Several definitions of a transmission line have been proposed. For our purposes a broad definition will be used: 'A transmission line is a system by which electromagnetic energy is guided from one point to another' [8]. This definition includes as transmission lines coaxial lines, open wire lines of one or more conductors, waveguides of various cross-section, and multiconductor cables.

1.2 Impedance concepts – lumped circuits

It follows from network theory that the characteristics of a linear network when examined from any pair of terminals at a given frequency, and under steady-state conditions, are indistinguishable from the characteristics of a simple combination (series or shunt) of a resistor* and an inductor or capacitor.

A voltage varying harmonically with frequency $f = \omega/2\pi$ can be represented by

* Energy sources are characterized by a negative resistor, and energy sinks by a positive resistor in their terminal impedances.

the complex function† $V \exp(\mathrm{j}\omega t)$. When this voltage is applied to the terminals of a linear network, the ratio of the voltage to the current into the terminals is, by definition, the impedance Z of the network at the terminals. For equivalent circuits of resistance R in series with inductance L or capacitance C,

$$Z = R + \mathrm{j}X \tag{1.1}$$

where $X = \omega L$ or $X = -1/\omega C$. For resistance R in parallel with inductance L or capacitance C,

$$Z = \frac{\mathrm{j}\omega LR}{R + \mathrm{j}\omega L}, \qquad Z = \frac{R}{1 + \mathrm{j}\omega CR} \tag{1.2}$$

respectively, In these latter cases it is more convenient to express the impedance in terms of its inverse, the admittance Y, writing

$$Y = G + \mathrm{j}B \tag{1.3}$$

where $G = 1/R$, and $B = -1/\omega L$ or $B = \omega C$.

When the impedance of a nonlinear network, or a network element such as a semiconductor diode, is given, it is understood to be the time average of the instantaneous impedance over a period of the applied voltage.

1.3 Impedance concepts – two-conductor transmission line

Consider a two-conductor transmission line, characterized by a distributed series inductance, a distributed series resistance, and with distributed shunt capacitance and conductance denoted by L, R, C and G, respectively, where these quantities are given in their appropriate units per metre. For an elemental section of this line, the voltage and current relations may be written as

$$\begin{aligned} \frac{\mathrm{d}V}{\mathrm{d}z} &= -\left(RI + L\frac{\mathrm{d}I}{\mathrm{d}t}\right) \\ \frac{\mathrm{d}I}{\mathrm{d}z} &= -\left(GV + C\frac{\mathrm{d}V}{\mathrm{d}t}\right) \end{aligned} \tag{1.4}$$

where V and I are the voltage and current at distance z along the line.* V and I are complex quantities whose time variation $\exp(\mathrm{j}\omega t)$ has been cancelled throughout eqn. 1.4. The general solution of eqn. 1.4 is

$$V = a \exp(-\gamma z) + b \exp(\gamma z) \tag{1.5}$$

† Actual, physical instantaneous voltages and currents are, of course, not complex, but may be regarded as the real, or the imaginary, or indeed the sum of these parts of complex functions.
* Eqns. 1.4 are approximations, usable for frequencies where the separation of the conductors is much less than the wavelength [9].

$$I = [a \exp(-\gamma z) - b \exp(\gamma z)]/Z_0 \tag{1.5}$$

where

$$Z_0 = \sqrt{\left(\frac{R + j\omega L}{G + j\omega C}\right)}, \qquad \gamma = \sqrt{[(R + j\omega L)(G + j\omega C)]} \tag{1.6}$$

a and b are constants determined by assigning specific values to V and I at some point along the line.

Z_0 is termed the 'characteristic impedance' and γ the 'propagation constant', and these may be written as $Z_0 = R_0 + jX_0\,[\Omega]$ and $\gamma = \alpha + j\beta[/\mathrm{m}]$, respectively, where α is the attenuation constant (nepers/m) and β the phase constant (radians/m). The eqns. 1.5 with the suppressed time factor $\exp(j\omega t)$ each represent two waves travelling in opposite directions, but having the same frequency. Each wave may be regarded as the summation of any number of waves of the same frequency having different amplitudes and phases. They are both periodic in the z direction. The spatial period, or wavelength λ, of a monochromatic wave is the distance between points of corresponding phase of consecutive cycles, and is therefore given by

$$\beta\lambda = 2\pi \tag{1.7}$$

from which

$$\lambda = 2\pi/\beta \tag{1.8}$$

The wave propagates with velocity v, where

$$v = f\lambda = \frac{2\pi f}{\beta} = \frac{\omega}{\beta} \tag{1.9}$$

The wave with coefficient a is exponentially attenuated in the positive z direction, and we shall call it the forward wave. The wave with coefficient b is exponentially attenuated in the negative z direction, and we shall call it the reflected wave.

For the special case of a lossless line with $R = G = 0$, which is a useful approximation to a practical low-loss line,

$$Z_0 = \sqrt{\frac{L}{C}}$$

$$\gamma = j\beta = j\omega\sqrt{(LC)}$$

$$\alpha = 0 \tag{1.10}$$

Since, in this case, the velocity of propagation is frequency independent, complex (multichromatic) signals propagating along a lossless twin wire transmission line arrive undistorted. The line is said to be non-dispersive, or distortionless.

More generally, the condition $\beta \propto \omega$, necessary for distortionless transmission, is seen from eqn. 1.6 to be satisfied provided $R/L = G/C$.

By applying the definition of impedance to the line at point z, we find from eqns. 1.5 that the impedance Z_{in} looking in the z direction is

$$Z_{in} = Z_0 \frac{a \exp(-\gamma z) + b \exp(\gamma z)}{a \exp(-\gamma z) - b \exp(\gamma z)} \tag{1.11}$$

If $b = 0$ there is no reflected wave, and $Z_{in} = Z_0$. Then the impedance everywhere along the line is the characteristic impedance, and the line is said to be matched. If the impedance anywhere along the (uniform) line is not equal to Z_0 the line is mismatched.

In order to find a relationship between the coefficients a and b, we cut the line at $z = 0$ and terminate it there with a load impedance Z_L, as shown in Fig. 1.1. Then from eqn. 1.11

$$Z_L = Z_0 \frac{a + b}{a - b} \tag{1.12}$$

and after rearranging:

$$\frac{b}{a} = \frac{Z_L - Z_0}{Z_L + Z_0} = \Gamma_L = |\Gamma_L| \exp(j\phi_L) \tag{1.13}$$

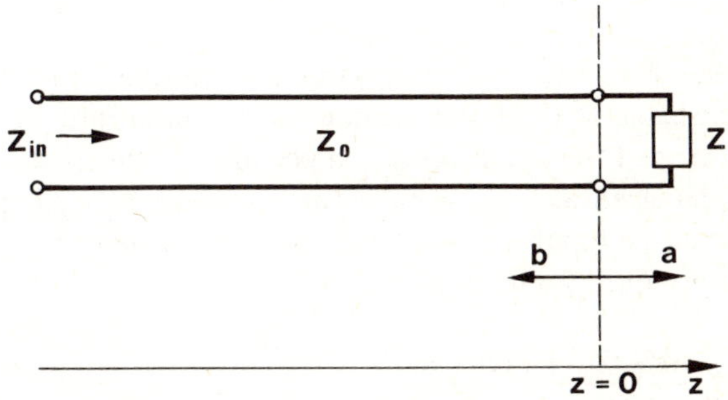

Fig. 1.1 *Uniform transmission line of characteristic impedance Z_0 terminated with load impedance Z_L at $z = 0$*

This complex ratio of the reflected to forward waves is termed the voltage reflection coefficient. Sometimes the symbol Γ_L is used for the negative of eqn. 1.13, which is the current reflection coefficient [10]. Throughout this book, reference to 'reflection coefficient' will be understood to mean voltage reflection coefficient.

That eqn. 1.13 defines a voltage reflection coefficient is seen by considering a transmission line terminated in a short circuit ($Z_L = 0$), across which no voltage can exist. Then eqn. 1.13 yields $\Gamma_L = -1$, which is the value required for the voltage of the reflected wave to cancel the voltage of the forward wave at the short circuit.

When Z_0 is real ($Z_0 = R_0$) it follows from eqn. 1.13 that $|\Gamma_L| \leqslant 1$ for all passive load impedances. The limiting case $|\Gamma_L| = 1$ is achieved whenever Z_L is imaginary.*

* If Z_0 is complex, it is possible for $|\Gamma_L| > 1$ when a passive load terminates the line. A necessary condition is that the imaginary parts of Z_0 and Z_L be of opposite sign [11].

When the transmission line is lossless ($\alpha = 0$), it follows from eqns. 1.5 and 1.13 that

$$\frac{V}{a} = \exp(-j\beta z) + \Gamma_L \exp(j\beta z)$$

'Return loss', defined by $-20 \log_{10}(|\Gamma_L|)$, is the magnitude of the ratio of reflected to forward voltages in (positive) decibels.

1.3.1 Impedance variation along a transmission line

We have already seen that if there is no reflected wave on the line, the impedance at any point is Z_0. This is true of an infinite line, and also for a line which is sufficiently long for voltages and currents to be measured before any reflected wave arrives.

If the terminated line in Fig. 1.1 is cut at a distance $l = -z$ from the termination to create a pair of input terminals, the input impedance Z_{in} is found from eqns. 1.11 and 1.13 to be

$$Z_{in} = Z_0 \frac{Z_L/Z_0 + \tanh(\gamma l)}{1 + Z_L/Z_0 \tanh(\gamma l)} \tag{1.14}$$

When $Z_L = Z_0$, then $Z_{in} = Z_0$ for any length l, and the line is therefore reflectionless and indistinguishable at the input terminals from an infinite line. It is for this reason that Z_0 is termed the characteristic impedance for the mode of propagation of interest, since the eigenvalue $Z_L = Z_0$ is the only terminating impedance which appears unchanged at the input terminals, regardless of line length.

For a lossless line, eqn. 1.14 reduces to

$$Z_{in} = Z_0 \frac{Z_L/Z_0 + j\tan(\beta l)}{1 + j(Z_L/Z_0)\tan(\beta l)} \tag{1.15}$$

Z_{in} is periodic with βl and remains unchanged if the line is lengthened or shortened in increments of $\lambda/2$. In particular, for $l = (2n+1)\lambda/4$, n an integer, $Z_{in} = Z_0^2/Z_L$, which equates the characteristic impedance to the geometric mean of the input and terminating impedances. (This relationship forms the basis for matching two different impedances using a quarter-wave transformer.) Similarly, if $l = n\lambda/2$, then $Z_{in} = Z_L$, and there is no impedance transformation between the termination and the input. (A multiple half-wave line is therefore useful to access a network port without impedance transformation.)

If the lossless line is terminated in a short circuit, $Z_{in} = jZ_0 \tan(\beta l)$, which for real Z_0 and $l < \lambda/4$ is imaginary and positive. Such an impedance is termed inductive, from a comparison with eqn. 1.1. Similarly, if the termination is an open circuit, $Z_{in} = -jZ_0 \cot(\beta l)$, which for real Z_0 and $l < \lambda/4$ is capacitive.

Although eqn. 1.14 for the lossy line and eqn. 1.15 for the lossless line describe the impedance transformation along the line, a simpler relationship may be found in terms of terminating and input reflection coefficients, instead of impedances.

Γ_{in} may be defined similarly to Γ_L in eqn. 1.13 by

$$\Gamma_{in} = \frac{Z_{in} - Z_0}{Z_{in} + Z_0} \tag{1.16}$$

and is the reflection coefficient, referenced to Z_0, at the input plane $l = -z$ of the transmission line.

From eqns. 1.13 and 1.16 we obtain

$$Z_{in} = Z_0 \frac{1 + \Gamma_{in}}{1 - \Gamma_{in}}, \qquad Z_L = Z_0 \frac{1 + \Gamma_L}{1 - \Gamma_L} \tag{1.17}$$

which, when substituted into eqn. 1.14, gives, for a lossy line,

$$\Gamma_{in} = \Gamma_L \exp(-2\gamma l) = \Gamma_L \exp(-2\alpha l) \exp(-j2\beta l) \tag{1.18}$$

This indicates that for a given terminating reflection coefficient Γ_L, the reflection coefficient Γ_{in} measured along the line will traverse an exponentially shrinking spiral on the complex Γ plane, and approach zero as distance increases from the load. As Γ_{in} approaches zero, Z_{in} approaches Z_0.

If the line is lossless ($\alpha = 0$)

$$\Gamma_{in} = \Gamma_L \exp(-j2\beta l) \tag{1.19}$$

This is a function which is periodic with $l = \lambda/2$, and on the complex Γ plane describes a circle with centre at the origin and radius $|\Gamma_L|$.

After comparing the complexities of eqn. 1.14 with eqn. 1.18 for the lossy case, and eqn. 1.15 with eqn. 1.19 for the lossless case, it is understandable why the use of the reflection coefficient has been preferred increasingly over the years to the use of impedance, when describing propagation along transmission lines.

1.3.2 Voltage variation along a transmission line

Consider a uniform transmission line, terminated at $z = 0$ by a load impedance Z_L. Without loss of generality, the voltage V along the line as given by eqn. 1.5 can be normalized by setting $a = 1$. This normalization is convenient as it both refers the phase of V, and scales the magnitude of V at any point along the line, to that of the forward wave incident on the termination. Then from eqn. 1.13, $\Gamma_L = b$, and

$$V = \exp(-\alpha z)\{\exp(-j\beta z) + |\Gamma_L| \exp(2\alpha z + j[\beta z + \phi_L])\} \tag{1.20}$$

For a lossless line, ($\alpha = 0$)

$$V = \exp(j\phi_L/2)\{(1 + |\Gamma_L|) \cos(\beta z + \phi_L/2) - j(1 - |\Gamma_L|) \sin(\beta z + \phi_L/2)\} \tag{1.21}$$

As z is varied, this expression describes an ellipse on the complex voltage plane. The ellipse has a semi-major axis of $1 + |\Gamma_L|$, a semi-minor axis of $1 - |\Gamma_L|$, and a rotation of the major axis of $\phi_L/2$ from the real axis of the complex plane, as shown in Fig. 1.2.

The voltage along the line may be expressed in polar co-ordinates as

$$V = |V| \exp (j\phi_v) \qquad (1.22)$$

where, from eqn. 1.21,

$$\tan\left(\beta z + \frac{\phi_L}{2}\right) \qquad (1.23)$$

$$\phi_v = \frac{\phi_L}{2} - \tan^{-1}\left[\frac{(1-|\Gamma_L|)}{(1+|\Gamma_L|)} \tan\left(\beta z + \frac{\phi_L}{2}\right)\right] \qquad (1.24)$$

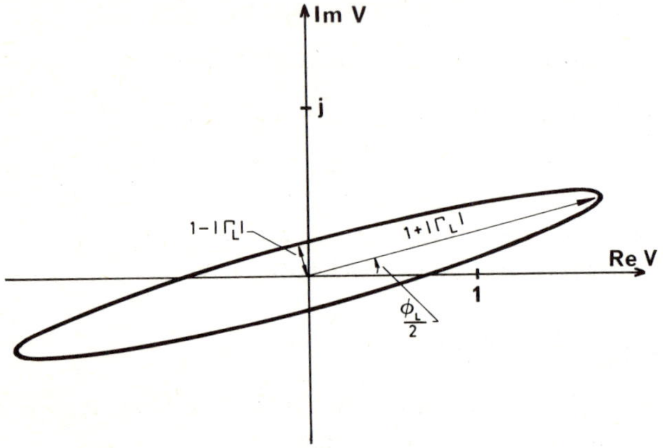

Fig. 1.2 *Locus of the complex voltage along a uniform transmission line terminated in an impedance having a reflection coefficient Γ_L. This locus is an ellipse with semi-major axis of $1 + |\Gamma_L|$, semi-minor axis of $1 - |\Gamma_L|$, and has a rotation of the major axis of $\phi_L/2$*

Eqn. 1.23 has great practical importance because it provides a means of determining both the magnitude and phase of Γ_L by making voltage measurements (magnitude only) along the line.

In particular, the voltage standing wave ratio (VSWR) denoted by the symbol S is, from eqn. 1.23,

$$S = \frac{|V|_{max}}{|V|_{min}} = \frac{1+|\Gamma_L|}{1-|\Gamma_L|} \qquad (1.25)$$

and therefore

$$|\Gamma_L| = \frac{S-1}{S+1} \qquad (1.26)$$

It follows from eqn. 1.23 that V has maxima when $2\beta z + \phi_L = -2n\pi$ ($n = 0$, 1, 2, . . .) and minima when $2\beta z + \phi_L = -(2n-1)\pi$, ($n = 0, 1, 2, \ldots$).

Thus the locations of the maxima or minima on the line provide the means of determining ϕ_L. The maxima occur when the voltages of the forward and reflected waves are in phase; the minima occur when the voltages are in anti-phase, a quarter-wavelength away from the maxima. The variations of both the magnitude and phase of voltage along a loss-free transmission line are illustrated in Figs. 1.3 and 1.4 for three values of $|\Gamma_L|$.

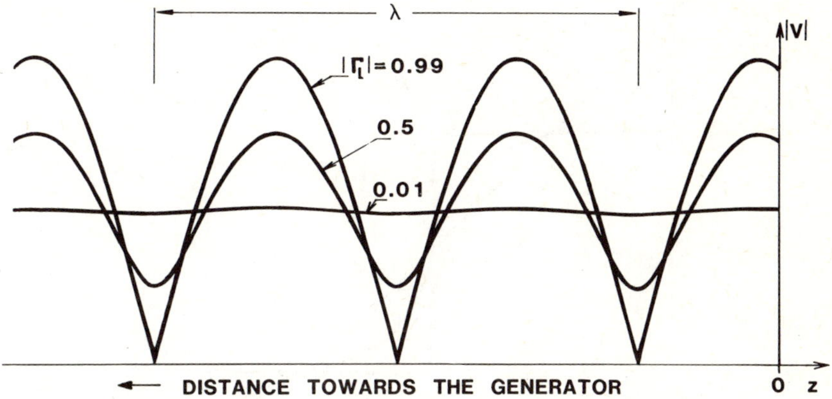

Fig. 1.3 *Magnitude of the voltage along a uniform lossless transmission line as a function of distance, for three different values of* $|\Gamma_L|$

For a lossy transmission line, eqn. 1.20 may be written as

$$V = \exp(-\alpha z + j\phi_L/2)[(1+r)\cos(\beta z + \phi_L/2) - j(1-r)\sin(\beta z + \phi_L/2)] \quad (1.27)$$

where $r = |\Gamma_L| \exp(2\alpha z)$. Thus the voltage encountered in moving from the load towards the generator (in the negative z direction) describes, on the complex voltage plane, an elliptical spiral with continually reducing ellipticity, as shown in Fig. 1.5. When $r \ll 1$ and can be neglected, the locus is an exponential spiral, which leads to the previous conclusion that a long lossy line behaves at its input as if it were terminated in its characteristic impedance.

1.3.3 Coaxial lines

Coaxial lines and cables are by far the most commonly used two-conductor transmission lines at microwave frequencies. Adequately terminated microwave coaxial lines using a cylinder of high conductivity as the outer conductor do not radiate, nor are they susceptible to external fields. When, for reasons of flexibility, the outer conductor is made as a tightly woven braid of high conductivity wire, the radiation is negligible and the cable is immune to external fields in all but the most demanding environments.

The coaxial cable with cross-section shown in Fig. 1.6*a* has an inner conductor of radius r_2, and outer conductor with (inner) radius r_1, and is filled with a homogeneous material of permeability μ and permittivity ϵ. The distributed shunt capacitance C is

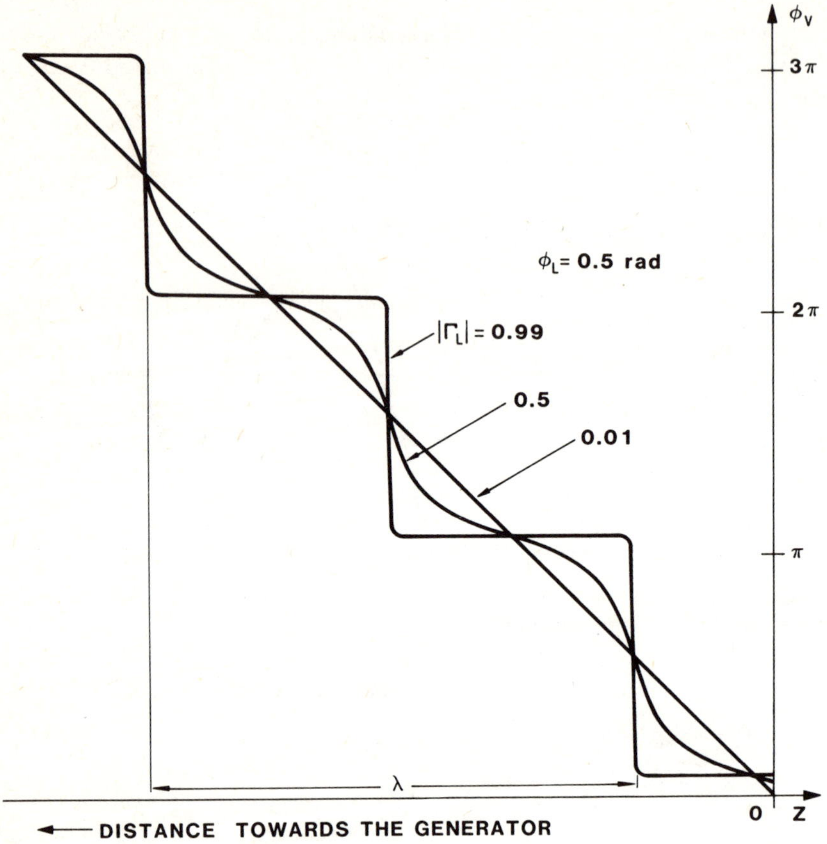

Fig. 1.4. *Phase of the voltage along a uniform lossless transmission line as a function of distance, for three different values of* $|\Gamma_L|$

$$C = \frac{2\pi\epsilon'}{\ln\left(\frac{r_1}{r_2}\right)} \qquad \text{F/m} \tag{1.28}$$

and the distributed shunt conductance G is

$$G = \frac{2\pi\omega\epsilon''}{\ln\left(\frac{r_1}{r_2}\right)} \qquad \text{S/m} \tag{1.29}$$

where $\epsilon = \epsilon' - j\epsilon''$. The series inductance L due to fields external to the conductors is

$$L = \frac{\mu}{2\pi} \ln\left(\frac{r_1}{r_2}\right) \qquad \text{H/m} \tag{1.30}$$

At high frequencies, skin effects resulting from the finite conductivity of the

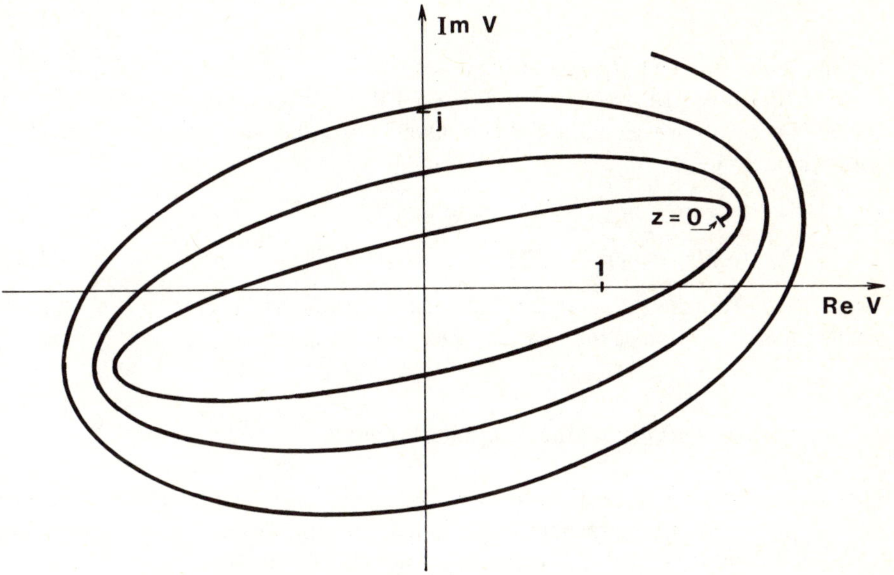

Fig. 1.5. *Locus of the complex voltage along a uniform lossy mismatched transmission line. This locus is an elliptic spiral*

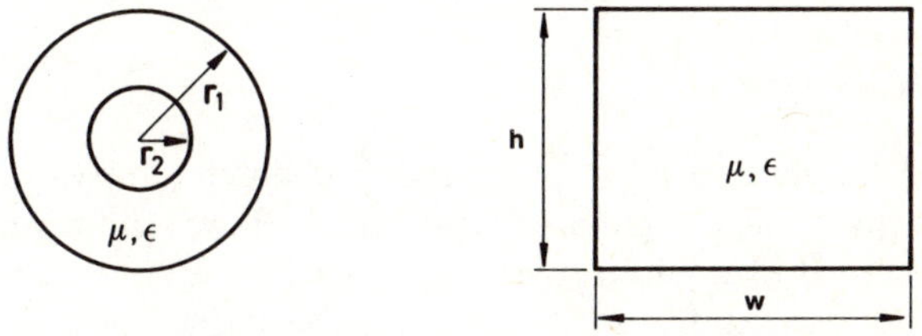

(a) (b)

Fig. 1.6. *Cross-sectional geometries of*
a coaxial transmission line
b rectangular waveguide

conductors introduce series resistance R_s, and, because of flux linkages within the conductors, additional series inductance L_s. This additional inductance may either be added directly to L, or because it has reactance equal to R_s, be combined conveniently with R_s to give a series impedance Z_s. Such series losses are discussed in more detail in Section 3.2.5.

From eqn. 1.6, the characteristic impedance and phase constant of a lossless coaxial line are

$$Z_0 = \frac{1}{2\pi}\sqrt{\frac{\mu}{\epsilon}}\ln\left(\frac{r_1}{r_2}\right) \tag{1.31}$$

$$\beta = \omega\sqrt{(\mu\epsilon)} \tag{1.31}$$

for the fundamental TEM propagating mode.

At sufficiently high frequencies, TE and TM modes can propagate along coaxial cable. The cutoff wavelengths for the lowest order TE and TM modes are given approximately by

$$\lambda_c \simeq \pi(r_1 + r_2) \qquad \text{TE mode}$$

$$\lambda_c \simeq 2(r_1 - r_2) \qquad \text{TM mode} \tag{1.32}$$

respectively. To ensure that these modes cannot propagate, coaxial cables are made smaller in cross-section as the intended frequency of use rises.

1.4 Impedance concepts in other transmission lines

The concepts of characteristic impedance, reflection coefficient, and standing wave ratio apply with varying degrees of usefulness to transmission systems supporting waveguide modes, as opposed to the fundamental TEM mode of propagation. For TEM mode propagation, identical results are obtained from the three expressions

$$Z = V/I$$

$$Z = W/II^* \tag{1.33}$$

$$Z = VV^*/W,$$

where W is the transmitted power, the asterisk denotes complex conjugation, and the values of voltage and current are unambiguous. However, for waveguide modes, in any cross-sectional plane the voltage varies between different points on the conductors, and the current has a non-constant distribution around the conductor surface. As a consequence, the values of impedance given by eqn. 1.33 not only differ, but depend on the particular location and manner in whcih voltage and current are measured.

For waveguide mode propagation, it is more convenient to think in terms of field concepts rather than voltage or current. Wave impedance is defined by $Z = E_t/H_t$, where E_t is the electric field intensity and H_t is the magnetic field intensity transverse to the direction of propagation. In cylindrical waveguides, E_t and H_t are functions of position across the guide, but for both TE and TM modes in uniform cylindrical guides of any cross-section the wave impedance is constant over the cross-section, and depends only on the properties of the material filling the guide and the ratio of signal frequency to cutoff frequency [12]. The behaviour of E_t along a waveguide is identical to the behaviour of voltage along a two-conductor transmission line, and similarly the behaviour of H_t along a waveguide is identical to the behaviour of current along a two-conductor transmission line. Consequently, both the concepts of reflection coefficient and standing wave ratio are unchanged.

However, the wave impedances cannot be used with formulas such as eqn. 1.13

to determine the reflection coefficient when two waveguides are joined. For example, two sections of rectangular waveguide of the same width but different height have identical wave impedances when filled with the same material and supporting the TE_{10} mode of propagation. Yet when these guides are joined, a reflected wave is set up.

Nevertheless, knowledge of the reflection coefficient Γ is sufficient to determine the proportion of incident signal reflected, and the proportion of incident power transferred across a discontinuity. More generally, knowledge of the scattering parameters discussed in Section 2.1.1 is sufficient to fully describe the behaviour of a device in the system of measurement, without the need to assign a value to the system impedance. Consequently, most of the measurement techniques described in this book are designed to determine relative impedance, which is conveniently expressed in terms of reflection coefficient.

Whereas impedance determination requires the absolute measurement of two quantities (voltage and current, voltage and power, current and power, or electric and magnetic field intensities), the determination of reflection coefficient requires only the relative measurement of two values of one quantity.

1.4.1 Propagation in rectangular waveguide

Transverse magnetic modes propagating in the z direction in the rectangular waveguide of height h and width w in Fig. 1.6*b* have field components proportional to

$$
\begin{aligned}
E_z &= \sin(Ax)\sin(By) \\
E_x &= \frac{-\mathrm{j}\beta A}{D^2}\cos(Ax)\sin(By) \\
E_y &= \frac{-\mathrm{j}\beta B}{D^2}\sin(Ax)\cos(By) \\
H_z &= 0 \\
H_x &= \frac{\mathrm{j}\omega\epsilon B}{D^2}\sin(Ax)\cos(By) \\
H_y &= \frac{-\mathrm{j}\omega\epsilon A}{D^2}\cos(Ax)\sin(By)
\end{aligned}
\tag{1.34}
$$

where

$$A = \frac{m\pi}{w}, \qquad B = \frac{n\pi}{h}, \qquad D^2 = A^2 + B^2$$

$$\beta = \left[\omega^2\mu\epsilon - \left(\frac{m\pi}{w}\right)^2 - \left(\frac{n\pi}{h}\right)^2\right]^{1/2} \tag{1.35}$$

and m, n are integers. The phase factor $\exp(\mathrm{j}\omega t - \mathrm{j}\beta z)$ at position z along the guide has been omitted from all field components for clarity. It is immediately apparent that, as stated in Section 1.4, the ratio of transverse electric to transverse magnetic fields is constant across the guide.

Similarly, transverse electric modes propagating in the z direction have components proportional to

$$\begin{aligned}
H_z &= \cos(Ax)\cos(By)\\
H_x &= \frac{j\beta A}{D^2}\sin(Ax)\cos(By)\\
H_y &= \frac{j\beta B}{D^2}\cos(Ax)\sin(By)\\
E_z &= 0\\
E_x &= \frac{j\omega\mu B}{D^2}\cos(Ax)\sin(By)\\
E_y &= \frac{-j\omega\mu A}{D^2}\sin(Ax)\cos(By)
\end{aligned} \tag{1.36}$$

and again, the ratio of transverse electric to transverse magnetic fields is constant across the guide, but with a value different to the TM case.

For both TM and TE modes, the wavelength of propagation λ_g along the guide is found, using eqn. 1.35, to be

$$\lambda_g = \frac{2\pi}{\beta} = \frac{2\pi}{\left[\omega^2\mu\epsilon - \left(\frac{m\pi}{w}\right)^2 - \left(\frac{n\pi}{h}\right)^2\right]^{1/2}} \tag{1.37}$$

The cutoff frequency, below which a mode does not propagate, is found by setting $\beta = 0$, and given by

$$f_c = \frac{1}{2(\mu\epsilon)^{1/2}}\left[\left(\frac{m}{w}\right)^2 + \left(\frac{n}{h}\right)^2\right]^{1/2} \tag{1.38}$$

and β, from eqn. 1.35 may be written as

$$\beta = \omega(\mu\epsilon)^{1/2}\left[1 - \left(\frac{f_c}{f}\right)^2\right]^{1/2} \tag{1.39}$$

Since β is not linearly dependent on frequency, propagation in rectangular waveguide is dispersive, and more so the nearer the frequency approaches the cutoff value.

The wavelength λ_c associated with the cutoff frequency, and corresponding to the wavelength of unguided propagation at a frequency f_c through the same material as fills the waveguide, is

$$\lambda_c = \frac{v}{f_c} = \frac{1}{f_c(\mu\epsilon)^{1/2}} = \frac{2}{\left[\left(\frac{m}{w}\right)^2 + \left(\frac{n}{h}\right)^2\right]^{1/2}} \tag{1.40}$$

Since the wavelength λ of unguided propagation at frequency f through the same material is $\lambda = 1/[f(\mu\epsilon)^{1/2}]$, the guide wavelength can be expressed as

$$\frac{1}{\lambda_g^2} = \frac{1}{\lambda^2} - \frac{1}{\lambda_c^2} \tag{1.41}$$

The wave impedance Z_w^{TM} for TM modes is, from eqn. 1.34,

$$Z_w^{TM} = \frac{\beta}{\omega\epsilon} = \left(\frac{\mu}{\epsilon}\right)^{1/2}\left[1 - \left(\frac{f_c}{f}\right)^2\right]^{1/2} \tag{1.42}$$

and for TE modes is, from eqn. 1.36,

$$Z_w^{TE} = \frac{\omega\mu}{\beta} = \left(\frac{\mu}{\epsilon}\right)^{1/2} \Big/ \left[1 - \left(\frac{f_c}{f}\right)^2\right]^{1/2} \tag{1.43}$$

When $w > h$, the lowest cutoff frequency is exhibited by the TE mode with $m = 1$, $n = 0$ (designated the TE_{10} mode), for which

$$\lambda_c(TE_{10}) = 2w \tag{1.44}$$

Consequently, for this, the fundamental mode, the guide wavelength is, from eqn. 1.37

$$\lambda_g = \frac{2w\lambda}{[(2w)^2 - \lambda^2]^{1/2}} \tag{1.45}$$

The guide wavelength, λ_g, is the length of one period of the spatial pattern of the *E*-field, *H*-field, and the wall currents along the guide. During one cycle, this pattern moves a distance λ_g, and therefore travels with a velocity, termed the 'phase velocity', of

$$v_p = f\lambda_g = f\frac{2\pi}{\beta} = \frac{\omega}{\beta} \tag{1.46}$$

However the phase velocity is not the velocity with which *information* is transmitted along dispersive lines. To convey information, changes must be made to the single frequency carrier signal, and these changes, termed 'modulation', require a 'group' of frequencies in the neighbourhood of the carrier to be propagated. The velocity with which the modulation is propagated is termed the 'group velocity', and describes the velocity with which information can be carried along the line.

The group velocity may be found conveniently by calculating the time of travel of a (sinc) pulse modulated carrier sent down a dispersive line.* However, since

* A sinc pulse has a constant magnitude band-limited spectrum.

each spectral component of the signal is subject to a different delay, the emergent pulse modulation will have a different shape to that transmitted (it will be dispersed), and it is therefore difficult to state when the pulse arrived. To make it easier to compare the exact temporal locations of pulses of different shapes, we can calculate the delay of increasingly narrow band pulse modulation, and imagine that in the limit, the required bandwidth tends to zero. Such pulses would, of course, have a broad main lobe (Chapter 6), but will arrive almost undistorted.

Assume that the dispersive line is lossless, with a phase constant β which can be approximated over the narrow band $\omega_0 \pm \delta$ of the pulse modulation by

$$\beta \simeq \beta_0 + m(\omega - \omega_0) \tag{1.47}$$

where

$$m = \left.\frac{\partial \beta}{\partial \omega}\right|_{\omega_0} \tag{1.48}$$

The emergent signal is calculable as the Fourier transform of the product of the pulse spectrum and the transfer function of the line (Chapter 6). This product is

$$\begin{aligned} T(\omega) &= e^{-j(\beta_0 - m\omega_0 + m\omega)l} = e^{-j\theta} e^{-jm\omega l}, \quad \omega_0 - \delta \leqslant |\omega| \leqslant \omega_0 + \delta \\ &= 0, \text{ elsewhere} \end{aligned} \tag{1.49}$$

and the Fourier transform is

$$\begin{aligned} f(t) &= \frac{1}{2\pi} \int_{-\infty}^{\infty} T(\omega) e^{j\omega t} d\omega \\ &= \frac{2\delta}{\pi} e^{-j\theta} \cos\,[(ml - t)\omega_0]\ \text{sinc}\left[\frac{(ml - t)\delta}{\pi}\right] \end{aligned} \tag{1.50}$$

where l is the line length. It can be seen that the arrival of the sinc pulse is delayed by a time ml. The group velocity, defined in the limit as the bandwidth approaches zero, is therefore

$$v_g = \frac{l}{ml} = \frac{\partial \omega}{\partial \beta} \tag{1.51}$$

Whereas the phase velocity is an easily measurable propagation parameter, the accurate measurement of group velocity is difficult, as it is only defined over an infinitesimal bandwidth, which cannot carry modulation.

For dispersionless lines, where $\beta \propto \omega$, phase and group velocities are identical since

$$\frac{\omega}{\beta} = \frac{\partial \omega}{\partial \beta} \tag{1.52}$$

References

1 Scientific Encyclopedia (Van Nostrand, 1968), p. 896

2 STEINMETZ, C.P.: 'Theory and calculation of alternating current phenomena' (W.J. Johnston, 1897), preface and p. 39

3 POHL, R.W.: 'Electricity and magnetism' (Blackie & Son, 1933), p. 174
4 HUTCHINSON, R.W.: 'Intermediate electricity' (University Tutorial Press, 1941), p. 519
5 SCHELKUNOFF, S.A.: 'The impedance concept and its applications to problems of reflection, reflection shielding and power absorption', *Bell Syst. Tech. J.*, 1938, **17,** p. 25
6 BODE, H.W.: 'Network analysis and feedback amplifier design' (Van Nostrand 1945, reprint 1957), p. 15
7 WARNER, F.L.: 'Microwave attenuation measurement' (Peter Peregrinus, 1977)
8 Scientific Encyclopedia (Van Nostrand, 1968), p. 1883
9 GABRIEL, J.G.: 'Theory of electromagnetic transmission structures Part I: relativistic foundations and network formalisms', *Proc. IEEE,* 1980, **68,** pp. 354–366
10 SLATER, J.C.: 'Microwave transmission' (McGraw-Hill, 1942), p. 25
11 SOMLO, P.I. and HUNTER, J.D.: 'Condition for reflection coefficient magnitude greater than one for passive transmission line and passive load', *IEEE Trans.*, 1981, **IM-30**, pp. 230–231
12 RAMO, S., WHINNERY, J.R. and VAN DUZER, T.: 'Fields and waves in communication electronics' (Wiley, 1965)

Chapter 2

Representation of networks

In order to analyse microwave impedance measurement techniques, methods are needed for representing, and if necessary, reducing the networks used to simpler configurations. Although a linear two-terminal (one-port) network may be fully described by its reflection coefficient normalized to a particular characteristic impedance, multi-terminal circuits require more parameters for their description. In this chapter some of these parameter families and their use will be discussed. The parameter families are, in general, a function of frequency.

2.1 Commonly used four-terminal parameters

A four terminal network is represented in Fig. 2.1, where the senses of the terminal voltages and currents are defined. The effect of embedding this network in a circuit is fully described by the interrelationships between these voltages and currents.

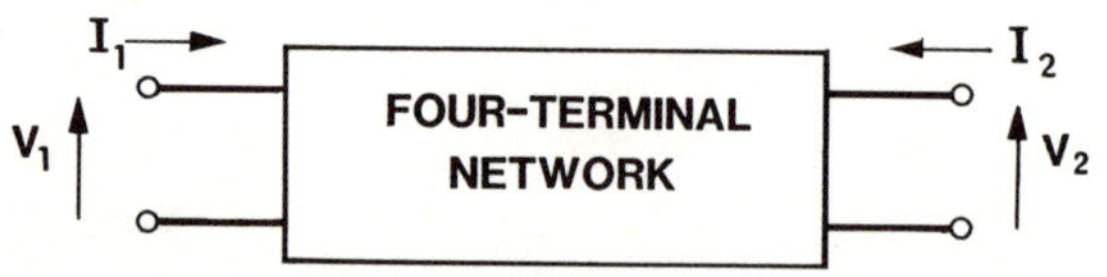

Fig. 2.1 *Definition of terminal voltages and currents for a four-terminal network*

If we express the left-hand (input) voltage and current in terms of the right-hand (output) voltage and current, then for linear networks we obtain the following linear characteristic equations:

$$\begin{aligned} V_1 &= AV_2 - BI_2 \\ I_1 &= CV_2 - DI_2 \end{aligned} \tag{2.1}$$

or in matrix form

$$\begin{bmatrix} V_1 \\ I_1 \end{bmatrix} = \begin{bmatrix} A & B \\ C & D \end{bmatrix} \begin{bmatrix} V_2 \\ -I_2 \end{bmatrix} \tag{2.2}$$

The parameters A, B, C and D have been variously called the asymmetrical linear parameters, the chain parameters, the cascading parameters, or simply the *ABCD* parameters, and form the elements of the *ABCD* or chain matrix. The negative signs in eqn. 2.1 are the consequence of the choice of the output current directions. It is evident from eqn. 2.1 that A and D are dimensionless, whereas B is dimensionally an impedance, and C is an admittance. The characteristic equations suggest a simple method for determining the individual parameters. If the output is short circuited, $V_2 = 0$, in which case $D = -I_1/I_2$ and $B = -V_1/I_2$. If the output is open circuited, $I_2 = 0$, and then $C = I_1/V_2$ and $A = V_1/V_2$. The parameters are, in general, complex and may be interpreted as

A: reverse voltage gain
$-B$: reverse transfer impedance
$-C$: reverse transfer admittance
D: reverse current gain

If the network is reciprocal (if a voltage V at one point in the network produces a current I at a second point in the network, then the same voltage V acting at the second point will produce current I at the first point), then the number of independent parameters reduces from four to three, by the link

$$AD - BC = 1 \tag{2.3}$$

Most passive circuits are reciprocal, with notable exceptions being those containing ferrites, in such components as isolators and circulators.

The terminal voltages may be expressed solely in terms of the terminal currents, using four coefficients with the dimensions of impedance. The characteristic impedance equations and the impedance matrix of the four-terminal are then

$$\begin{bmatrix} V_1 \\ V_2 \end{bmatrix} = \begin{bmatrix} Z_{11} & Z_{12} \\ Z_{21} & Z_{22} \end{bmatrix} \begin{bmatrix} I_1 \\ I_2 \end{bmatrix} \tag{2.4}$$

where, with reciprocity,

$$Z_{12} = Z_{21} \tag{2.5}$$

The dual of the impedance matrix is the admittance matrix with the characteristic equations

$$\begin{bmatrix} I_1 \\ I_2 \end{bmatrix} = \begin{bmatrix} Y_{11} & Y_{12} \\ Y_{21} & Y_{22} \end{bmatrix} \begin{bmatrix} V_1 \\ V_2 \end{bmatrix} \tag{2.6}$$

with the link for reciprocity

$$Y_{12} = Y_{21} \tag{2.7}$$

High frequency transistor performance is often defined by the 'h' parameters, or skew-symmetric parameters given by

$$\begin{bmatrix} V_1 \\ I_2 \end{bmatrix} = \begin{bmatrix} h_{11} & h_{12} \\ h_{21} & h_{22} \end{bmatrix} \begin{bmatrix} I_1 \\ V_2 \end{bmatrix} \tag{2.8}$$

with, for reciprocal networks

$$h_{12} = h_{21} \tag{2.9}$$

For all the above parameter families the senses of voltages and currents are as shown in Fig. 2.1.

The usefulness of these parameter families lies in the simplicity of the mathematical operations needed to describe the connection of two four-terminal networks in the various ways shown in Fig. 2.2. If two four-terminal networks are cascaded to form a chain (Fig. 2.2*a*), multiplying the chain (*ABCD*) matrices in the proper order yields the chain matrix of the resultant network. If we distinguish the chain parameters of the network nearer the source by the superscripts a, and those of the network further from the source by the superscript b, the chain matrix of the resultant network is

$$\begin{bmatrix} A^a & B^a \\ C^a & D^a \end{bmatrix} \begin{bmatrix} A^b & B^b \\ C^b & D^b \end{bmatrix} = \begin{bmatrix} A^aA^b + B^aC^b & A^aB^b + B^aD^b \\ C^aA^b + D^aC^b & C^aB^b + D^aD^b \end{bmatrix} \tag{2.10}$$

If the networks are connected in series (Fig. 2.2*b*), addition of the impedance matrices yields the resultant impedance matrix. Thus

$$\begin{bmatrix} Z_{11}^a & Z_{12}^a \\ Z_{21}^a & Z_{22}^a \end{bmatrix} + \begin{bmatrix} Z_{11}^b & Z_{12}^b \\ Z_{21}^b & Z_{22}^b \end{bmatrix} = \begin{bmatrix} Z_{11}^a + Z_{11}^b & Z_{12}^a + Z_{12}^b \\ Z_{21}^a + Z_{21}^b & Z_{22}^a + Z_{22}^b \end{bmatrix} \tag{2.11}$$

Similarly, when networks are paralleled (Fig. 2.2*c*) the admittance matrices are additive, and when networks are series-paralleled (Fig. 2.2*d*), the h parameters are additive.

There are, of course, more parameter families, such as the equivalent tee or pi network parameters, and the open and short circuit parameters. These families do not have characteristic equations, but are useful in special cases. At higher frequencies, where the direct measurement of voltages and currents is impracticable, the image parameters are sometimes used for reciprocal networks. For symmetrical networks, the image parameters degenerate into the equivalent transmission line parameters of characteristic impedance, propagation constant, and length.

The S, or scattering parameter family, is particularly useful for the analysis of microwave circuits, and will be discussed in detail in the next section. Related parameter families include the gathering and wave-cascading, or T coefficients [1].

2.1.1 S and T parameters

The scattering or S parameter family is designed to relate incoming and outgoing signals or waves at the ports of a network. In general, an N-port network has N^2 S parameters associated with N designated characteristic impedances Z_{0i}, $i = 1, 2, \ldots N$.

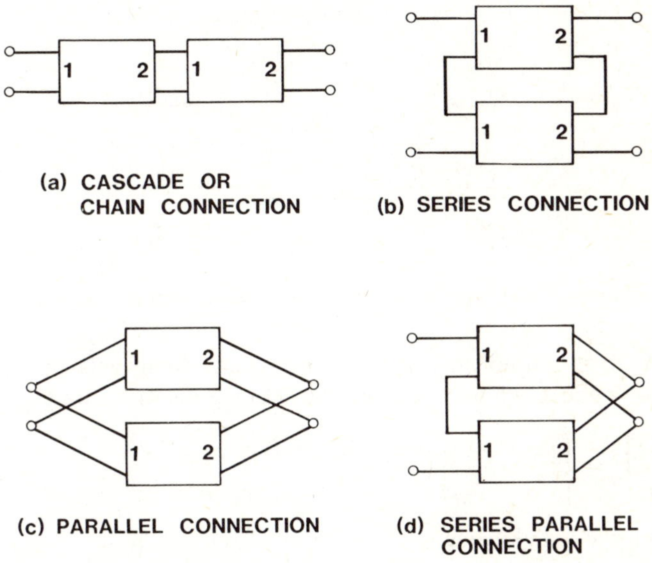

Fig. 2.2 *Various interconnections of two four-terminal networks*

The two-port in Fig. 2.3 is shown with terminal surfaces 1 and 2. In waveguide networks it is common, but not necessary, to choose these surfaces to coincide with the waveguide flanges. A signal incident onto a terminal surface from outside the network is represented by the letter a, and a signal leaving the terminal surface is represented by the letter b. In general, a and b are complex. The four solid lines in Fig. 2.3 represent these directional signals. Although the path of each pair of incoming and outgoing signals is in the same transmission line, the four lines should not be confused with the conductors shown in Figs. 2.1 and 2.2.

For a linear two-port network, the defining equations are

$$\begin{bmatrix} b_1 \\ b_2 \end{bmatrix} = \begin{bmatrix} S_{11} & S_{12} \\ S_{21} & S_{22} \end{bmatrix} \begin{bmatrix} a_1 \\ a_2 \end{bmatrix} \tag{2.12}$$

which emphasize that the outgoing signal at a port is linearly related to the incoming signals at all ports. The values of the S parameters depend on the designated characteristic impedances Z_{01} and Z_{02} associated with ports 1 and 2, respectively, and therefore the S parameters alone are not a complete description of the network characteristics. The reciprocity link is

$$S_{12}Z_{02} = S_{21}Z_{01} \tag{2.13}$$

The S parameters can be interpreted as reflection and transmission coefficients. If port 2 of the two-port is terminated in its designated impedance Z_{02} (a matched load in a Z_{02} system), $a_2 = 0$, and $S_{11} = b_1/a_1$, which is the reflection coefficient at port 1 in the designated Z_{01} impedance system. Similarly, S_{22} is the reflection coefficient at port 2 in a Z_{02} system, and S_{12}, S_{21} are transmission coefficients when ports 1 and 2 are terminated in their designated impedances.

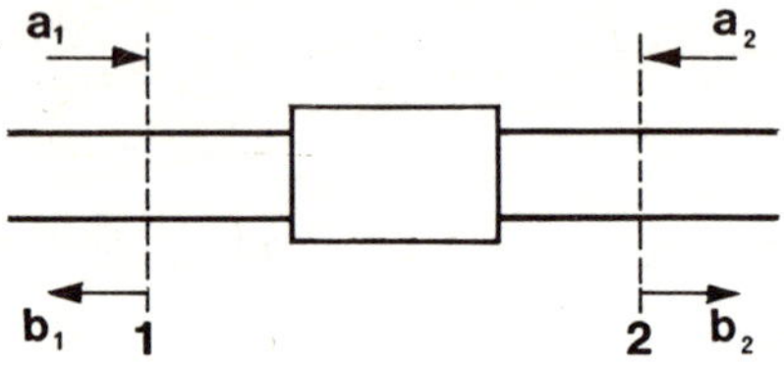

Fig. 2.3 *Definition of incident and emergent signals at the ports of a two-port network*

The cascading, transmitting, or T parameters are closely related to the S parameters, and for a linear two-port network the defining equations are [1]

$$\begin{bmatrix} b_1 \\ a_1 \end{bmatrix} = \begin{bmatrix} T_{11} & T_{12} \\ T_{21} & T_{22} \end{bmatrix} \begin{bmatrix} a_2 \\ b_2 \end{bmatrix} \tag{2.14}$$

with the reciprocity link being

$$T_{11}T_{22} - T_{12}T_{21} = \frac{Z_{01}}{Z_{02}} \tag{2.15}$$

When two networks are cascaded, the T matrix of the composite network is given by the product of the T matrices of the cascaded networks, taken in the proper order.

2.1.2 Matrix renormalization

Although the matrix S containing the S parameters of a network is associated with a set of characteristic impedances Z_{0i}, it is possible to transform S into a matrix S' associated with a different set of characteristic impedances ξ_{0i}. This transformation is known as matrix renormalization, and is performed using [2, 3]

$$S' = S - (I + S)\Gamma(I - S\Gamma)^{-1}(I - S) \tag{2.16}$$

where I is the identity matrix, and Γ is a diagonal matrix of reflection coefficients of the loads ξ_{0i} referred to the impedances Z_{0i}. Other forms of this expression are available, as are explicit expansions for 2, 3, 4, 5 and 6-port networks [4–6].

The availability of the matrix renormalization formula enables S, measured with arbitrary but known assigned terminations connected to each port, to be transformed into S', the S parameter set associated with the desired port impedances

ξ_{0i}. Terminations with the values ξ_{0i} are therefore not required during the measurement.

When a network with known S and Z_{0i} has ports terminated in other impedances ξ_{0i}, the application of matrix renormalization to find S' considerably simplifies the process of flow graph reduction discussed in later sections. For example, with terminations ξ_{0i} the transmission from port n to port m is a complicated function of the elements of S, but is simply element S'_{mn} of S'.

2.1.3 Relationships between parameter families

The relationships between various parameter families are given in Table 2.1 [7]. Neither reciprocity is assumed, nor is the designated impedance Z_{01} of port 1 required to equal the designated impedance Z_{02} of port 2.

If reciprocity is assumed, and $Z_{01} = Z_{02}$, the relationships in Table 2.1 reduce to those available in earlier literature [8–10].

Table 2.1 *Relationships between various two-port parameter families*

$$\begin{bmatrix} A & B \\ C & D \end{bmatrix} = \frac{1}{Z_{21}} \begin{bmatrix} Z_{11} & (Z_{11}Z_{22} - Z_{12}Z_{21}) \\ 1 & Z_{22} \end{bmatrix}$$

$$\begin{bmatrix} A & B \\ C & D \end{bmatrix} = \frac{-1}{Y_{21}} \begin{bmatrix} Y_{22} & 1 \\ (Y_{11}Y_{22} - Y_{12}Y_{21}) & Y_{11} \end{bmatrix}$$

$$\begin{bmatrix} A & B \\ C & D \end{bmatrix} = \frac{1}{2S_{21}} \begin{bmatrix} [(1+S_{11})(1-S_{22}) + S_{12}S_{21}] & Z_{02}[(1+S_{11})(1+S_{22}) - S_{12}S_{21}] \\ \dfrac{1}{Z_{01}}[(1-S_{11})(1-S_{22}) - S_{12}S_{21}] & \dfrac{Z_{01}}{Z_{02}}[(1-S_{11})(1+S_{22}) + S_{12}S_{21}] \end{bmatrix}$$

$$\begin{bmatrix} A & B \\ C & D \end{bmatrix} = \frac{1}{2} \begin{bmatrix} (T_{22} + T_{11} + T_{21} + T_{12}) & Z_{02}(T_{22} - T_{11} - T_{21} + T_{12}) \\ \dfrac{1}{Z_{01}}(T_{22} - T_{11} + T_{21} - T_{12}) & \dfrac{Z_{01}}{Z_{02}}(T_{22} + T_{11} - T_{21} - T_{12}) \end{bmatrix}$$

$$\begin{bmatrix} Z_{11} & Z_{12} \\ Z_{21} & Z_{22} \end{bmatrix} = \frac{1}{C} \begin{bmatrix} A & (AD - BC) \\ 1 & D \end{bmatrix}$$

$$\begin{bmatrix} Z_{11} & Z_{12} \\ Z_{21} & Z_{22} \end{bmatrix} = \frac{1}{Y_{11}Y_{22} - Y_{12}Y_{21}} \begin{bmatrix} Y_{22} & -Y_{12} \\ -Y_{21} & Y_{11} \end{bmatrix}$$

$$\begin{bmatrix} Z_{11} & Z_{12} \\ Z_{21} & Z_{22} \end{bmatrix} = \frac{1}{(1-S_{11})(1-S_{22}) - S_{12}S_{21}} \cdot$$

Table 2.1 *Cont.*

$$\begin{bmatrix} Z_{01}[(1+S_{11})(1-S_{22})+S_{12}S_{21}] & 2Z_{01}S_{21} \\ 2Z_{02}S_{12} & Z_{02}[(1-S_{11})(1+S_{22})+S_{12}S_{21}] \end{bmatrix}$$

$$\begin{bmatrix} Y_{11} & Y_{12} \\ Y_{21} & Y_{22} \end{bmatrix} = \frac{1}{B}\begin{bmatrix} D & -(AD-BC) \\ -1 & A \end{bmatrix}$$

$$\begin{bmatrix} Y_{11} & Y_{12} \\ Y_{21} & Y_{22} \end{bmatrix} = \frac{1}{Z_{11}Z_{22}-Z_{12}Z_{21}}\begin{bmatrix} Z_{22} & -Z_{12} \\ -Z_{21} & Z_{11} \end{bmatrix}$$

$$\begin{bmatrix} Y_{11} & Y_{12} \\ Y_{21} & Y_{22} \end{bmatrix} = \frac{1}{(1+S_{11})(1+S_{22})-S_{12}S_{21}} \cdot$$

$$\begin{bmatrix} Y_{01}[(1-S_{11})(1+S_{22})+S_{12}S_{21} & -2Y_{01}S_{12} \\ -2Y_{02}S_{21} & Y_{02}[(1+S_{11})(1-S_{22})+S_{12}S_{21}] \end{bmatrix}$$

$$\begin{bmatrix} S_{11} & S_{12} \\ S_{21} & S_{22} \end{bmatrix} = \frac{1}{(B+CZ_{01}Z_{02})+(AZ_{02}+DZ_{01})} \cdot$$

$$\begin{bmatrix} (B-CZ_{01}Z_{02}+AZ_{02}-DZ_{01}) & 2Z_{01}(AD-BC) \\ 2Z_{02} & (B-CZ_{01}Z_{02}-AZ_{02}+DZ_{01}) \end{bmatrix}$$

$$\begin{bmatrix} S_{11} & S_{12} \\ S_{21} & S_{22} \end{bmatrix} = \frac{1}{(Z_{11}+Z_{01})(Z_{22}+Z_{02})-Z_{12}Z_{21}} \cdot$$

$$\begin{bmatrix} [(Z_{11}-Z_{01})(Z_{22}+Z_{02})-Z_{12}Z_{21}] & 2Z_{01}Z_{12} \\ 2Z_{02}Z_{21} & [(Z_{11}+Z_{01})(Z_{22}-Z_{02})-Z_{12}Z_{21}] \end{bmatrix}$$

$$\begin{bmatrix} S_{11} & S_{12} \\ S_{21} & S_{22} \end{bmatrix} = \frac{-1}{(Y_{11}+Y_{01})(Y_{22}+Y_{02})-Y_{12}Y_{21}} \cdot$$

$$\begin{bmatrix} [(Y_{11}-Y_{01})(Y_{22}+Y_{02})-Y_{12}Y_{21}] & 2Y_{02}Y_{12} \\ 2Y_{01}Y_{21} & [(Y_{11}+Y_{01})(Y_{22}-Y_{02})-Y_{12}Y_{21}] \end{bmatrix}$$

$$\begin{bmatrix} S_{11} & S_{12} \\ S_{21} & S_{22} \end{bmatrix} = \frac{1}{T_{22}}\begin{bmatrix} T_{12} & (T_{11}T_{22}-T_{12}T_{21}) \\ 1 & -T_{21} \end{bmatrix}$$

$$\begin{bmatrix} T_{11} & T_{12} \\ T_{21} & T_{22} \end{bmatrix} = \frac{1}{S_{21}}\begin{bmatrix} (S_{12}S_{21}-S_{11}S_{22}) & S_{11} \\ -S_{22} & 1 \end{bmatrix}$$

2.2 Signal flow graphs

A signal flow graph is a symbolic, graphical way to represent simultaneous, linear equations. The flow graph consists of branches (directed paths) bearing values, and of junctions of paths, termed nodes. Each variable is represented by a node, and each branch is assigned the value of the coefficient linking the end variables.

Three linear equations and their flow graph representations are shown in Fig. 2.4, and the flow graphs representing linear source, load, and detector one-ports are shown in Fig. 2.5. It is convenient to represent a source as a constant generator of outward travelling waves, with a source reflection coefficient Γ_G looking back into the output port. This differs from Thévenin or Norton circuit models which represent the source either as a constant voltage generator with a series impedance, or a constant current generator with a shunt admittance [11].

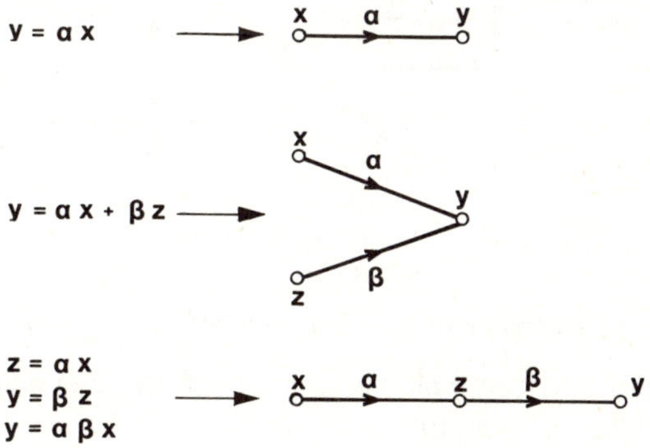

Fig. 2.4 *Three linear equations and their signal flow graph representations*

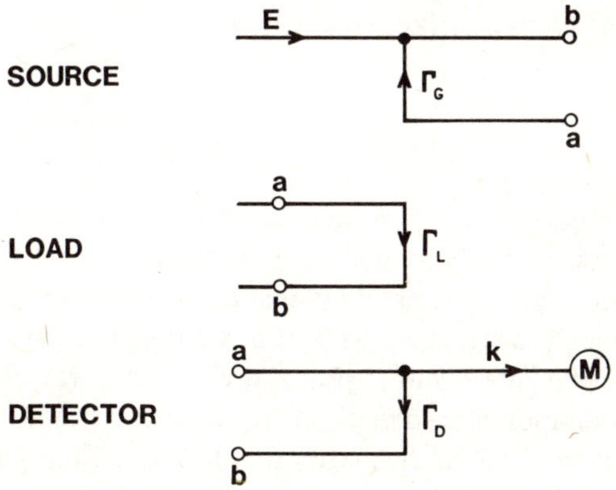

Fig. 2.5 *Flow graphs of three commonly used one-ports*

The flow graph of the two-port described by eqn. 2.12 is shown in Fig. 2.6. In particular, when the two-port is a series impedance Z,

$$S_{11} = S_{22} = \Gamma, \qquad S_{12} = S_{21} = 1 - \Gamma \tag{2.17}$$

where, from eqn. 1.16

$$\Gamma = \frac{Z}{Z + 2Z_0} \tag{2.18}$$

When the two-port is a shunt admittance Y,

$$S_{11} = S_{22} = \Gamma, \qquad S_{12} = S_{21} = 1 + \Gamma \tag{2.19}$$

with

$$\Gamma = \frac{-Y}{Y + 2Y_0} \tag{2.20}$$

The flow graph of an N-port is constructed by extending the two-port graph in Fig. 2.6 to have $2N$ nodes and N^2 branches.

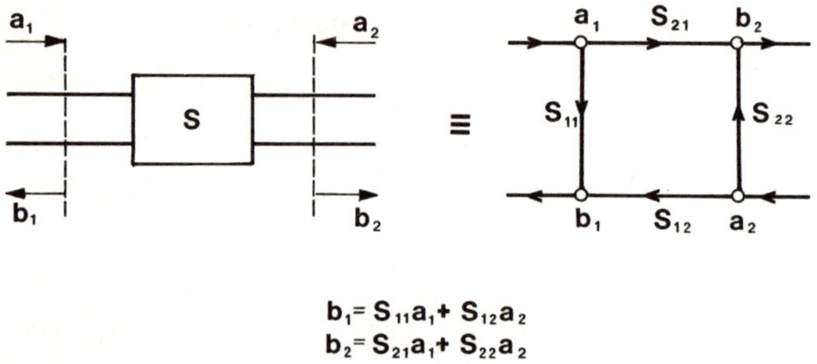

Fig. 2.6 *Flow graph representation of a one-port in terms of its S parameters*

Complicated linear microwave networks may be fully described by a set of linear equations, and the flow graphs for most of them can be constructed simply by connecting the flow graph representations of the network components. Any one of a number of flow graph reduction methods can then be used to determine the relationship between any two of the signal variables.

2.3 Flow graph reduction methods

The purpose of flow graph reduction methods is to reduce a generally complicated flow graph into a single branch connecting two variables. The value of the branch is then the ratio of the variables. The reduction methods provide answers to questions such as 'if a given signal enters port n, what signal will emerge at port m?', or 'what is the reflection coefficient at port i, if all other ports are short circuited?'.

There are three commonly used methods for reducing flow graphs: a topological method, an algebraic method, and a matrix method suitable for implementation on digital computers.

2.3.1 Topological flow graph reduction

This method, summarized by four simple rules given by Kuhn [11], reduces a complicated flow graph in a step-by-step procedure to successively simpler graphs, ending in a single path connecting two nodes. The four rules are:

I Branches in series multiply.

II Branches in parallel add. For parallelism the arrows must point in the same direction – if seemingly parallel branches have arrows pointing in opposite directions, a loop may be formed by duplicating a node according to Rule IV.

III A loop of value K attached to a node may be eliminated by dividing all incoming branches to that node by $1-K$.

IV A node may be duplicated provided that every path on the original graph is maintained. Loops attached to a duplicated node will also be duplicated.

The above rules are given here without proof, but their use is demonstrated in the example of Fig. 2.7 to find the input reflection coefficient of a terminated two-port. The initial flow graph is systematically reduced to a single path linking the incident and reflected signals.

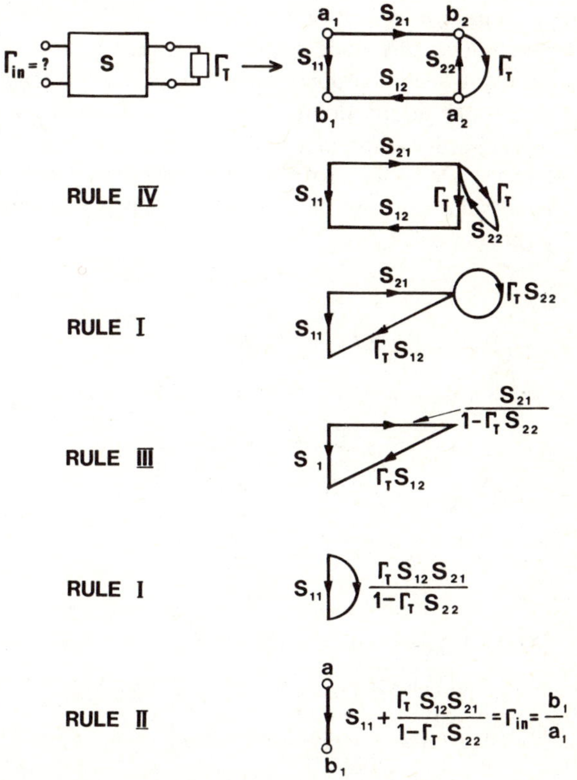

Fig. 2.7 *An example of the topological flow graph reduction technique using the rules given in Section 2.3.1*

2.3.2 Flow graph reduction with Mason's non-touching loop rule

Warner [12] has summarized Mason's rule [13, 14] which is

$$T = \frac{P_1(1-\Sigma^1 L_1 + \Sigma^1 L_2 - \ldots) + P_2(1-\Sigma^2 L_1 + \Sigma^2 L_2 - \ldots) + \ldots}{1-\Sigma L_1 + \Sigma L_2 - \ldots} \tag{2.21}$$

where T is the coefficient linking two variables ($b = Ta$) and where

P_1	is one path from a to b
P_2	is a different path from a to b, etc.
ΣL_1	is the sum of all 1st-order loops
ΣL_2	is the sum of all 2nd-order loops, etc.
$\Sigma^1 L_1$	is the sum of all 1st-order loops not touching P_1
$\Sigma^1 L_2$	is the sum of all 2nd-order loops not touching P_1
$\Sigma^2 L_1$	is the sum of all 1st-order loops not touching P_2
$\Sigma^2 L_2$	is the sum of all 2nd-order loops not touching P_2
etc.	

A path is the product of all branches encountered *en route* following the directions of arrows and not passing any node more than once. There may be many distinct paths from node a to node b. A first-order loop is a closed path in the directions of the arrows, not passing over any node more than once. Its value is the product of all branches traversed. A second-order loop is the product of any two first-order loops which do not touch at any point. A third-order loop is the product of any three non-touching first-order loops, and so on.

The use of Mason's rule is demonstrated using the flow graph in Fig. 2.8. The input reflection coefficient (at port 1) of a 3-port is calculated when ports 2 and 3 are terminated by short circuits. Then

$$
\begin{aligned}
P_1 &= S_{11} & \Sigma L_1 &= -S_{22} - S_{33} + S_{23}S_{32}\\
P_2 &= -S_{21}S_{12} & \Sigma L_2 &= S_{22}S_{33}\\
P_3 &= -S_{31}S_{13} & \Sigma^1 L_1 &= -S_{22} - S_{33} + S_{23}S_{32}\\
& & \Sigma^1 L_2 &= S_{22}S_{33}\\
P_4 &= S_{31}S_{23}S_{12} & \Sigma^2 L_1 &= -S_{33}\\
P_5 &= S_{21}S_{32}S_{13} & \Sigma^3 L_1 &= -S_{22}
\end{aligned}
$$

Applying eqn. 2.21,

$$
\begin{aligned}
\Gamma_{in} = {} & [S_{11}(1 + S_{22} + S_{33} - S_{23}S_{32} + S_{22}S_{33}) - S_{21}S_{12}(1 + S_{33})\\
& - S_{31}S_{13}(1 + S_{22}) + S_{31}S_{23}S_{12} + S_{21}S_{32}S_{13}]/\\
& [1 + S_{22} + S_{33} - S_{23}S_{32} + S_{22}S_{33}]
\end{aligned}
\tag{2.22}
$$

2.3.3 Matrix method of flow graph reduction

Munro and McMorran [15]* have described a very simple and effective method particularly suited for digital computers.

The N nodes in the flow graph are numbered from 1 to N. An $N \times N$ matrix is

* In reference 15 the rule of matrix reduction has errors in the wording.

formed with the branch values S_{ij} in the i, j position. If the desired answer is the ratio of the signal at the qth node to that at the pth node, the matrix is then augmented by an extra $(N+1)$th row and column containing zeros except in the positions $(q, N+1)$ and $(N+1, p)$, which contain values of 1. The augmented matrix is reduced in the following manner. For $j = 1, \ldots, N$, then for $i = j+1, \ldots N+1$, add $S_{ij}S_{jk}/(1-S_{jj})$ to S_{ik} for $k = j+1, \ldots, N+1$. The required answer is given by element $S_{N+1,N+1}$.

The original $N \times N$ submatrix can be reused to determine the transmission between other node pairs after relocating the unit entries in row $(N+1)$ and column $(N+1)$.

The implementation of the above reduction rules is straightforward on a digital computer, and a typical program is outlined in Appendix A.

One example of the usefulness of this matrix reduction technique is a computer program which has been written to analyse microwave networks.† The program contains a catalogue of S parameter matrices representing lumped series and shunt impedances, impedance transformers, variable lengths of transmission line, and more complicated components such as a slotted line and probe whose representation is based on two sections of transmission line whose lengths vary while the overall length remains constant. The program constructs the network of interest from these building blocks, and reduces the flow graph to find the required signal transfer.

It is helpful to have available a digital flow graph reduction program for checking the algebraic reduction of complicated flow graphs. The final algebraic expression can be compared numerically with the digital result after assigning arbitrary values to all branches.

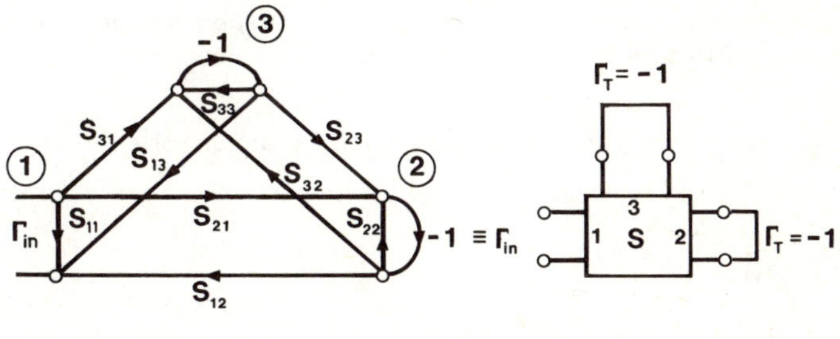

Fig. 2.8 *Flow graph representation of a general three-port network with short-circuit terminations on ports 2 and 3*

2.4 Two-port error box

It is helpful in many impedance measurement systems to represent a linear but imperfect measuring instrument as a perfect measuring instrument (core) connected

† P.I. Somlo

to the load impedance via a fictitious reciprocal two-port network (error box), as shown in Fig. 2.9. Once the two-port has been characterized, it provides the means for determining the reflection coefficient Γ_L of the load from the value Γ indicated by the measuring instrument.

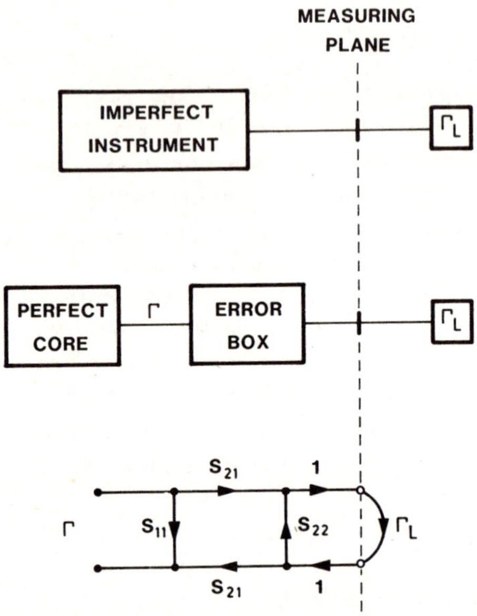

Fig. 2.9 *Linear but imperfect impedance measuring instrument represented as a perfect measuring instrument (core), with a fictitious two-port network (error box) which transforms Γ_L into Γ*

The value Γ sensed by the core and indicated by the imperfect measuring instrument is related to Γ_L by

$$\Gamma = S_{11} + \frac{S_{21}^2 \Gamma_L}{1 - \Gamma_L S_{22}} \tag{2.23}$$

which is recognizable as a bilinear transformation of the form

$$\Gamma = \frac{a + b\Gamma_L}{c + d\Gamma_L} \tag{2.24}$$

The parameters of the error box may be found by connecting successively three known terminations to the measuring port. When these terminations are a matched load, a short circuit, and an open circuit for which the reflection coefficients, as measured by the imperfect instrument, are Γ_m, Γ_s, Γ_o, respectively, then

$$S_{11} = \Gamma_m$$

$$S_{22} = -\frac{(\Gamma_s - \Gamma_m) + (\Gamma_o - \Gamma_m)}{\Gamma_s - \Gamma_o}$$

$$S_{21}^2 = \frac{2(\Gamma_s - \Gamma_m)(\Gamma_o - \Gamma_m)}{\Gamma_s - \Gamma_o} \tag{2.25}$$

Any subsequent measurement provides the value of the unknown Γ_L from Γ by rearranging eqn. 2.23 as

$$\Gamma_L = \frac{\Gamma - S_{11}}{S_{21}^2 + S_{22}(\Gamma - S_{11})} \tag{2.26}$$

The application of the error box concept is quite general, provided that the imperfect measuring apparatus responds linearly. Any non-linear effects caused by signal compression, interaction between magnitude and phase sensing, or level sensitivities of any other kind are not taken into account.

2.5 Graphical impedance representations

2.5.1 Cartesian and polar impedance planes

In general, impedances are represented by complex numbers, and therefore the plotting of impedances must be done on a complex plane. If an impedance is given by its real and imaginary parts, it may be plotted directly in a Cartesian co-ordinate system, where the two axes represent real and imaginary numbers. If the impedance is given by its magnitude and phase angle, a polar co-ordinate system is more convenient. Complex number addition may be visualized as a vectorial addition in a Cartesian co-ordinate system, since the sum of two complex numbers is obtained by adding their real and their imaginary parts. Multiplication is best visualized in a polar co-ordinate system, since the product of two complex numbers is obtained by multiplying their magnitudes, and adding arguments.

2.5.2 Smith chart

Smith [16, 17] introduced an alternative co-ordinate system which is particularly well-suited for plotting impedances in transmission line problems. If the impedances are normalized (usually to the characteristic impedance of the transmission line system in which the impedance is measured), a reflection coefficient may be defined from eqn. 1.13 as

$$\Gamma_L = \frac{Z' - 1}{Z' + 1} \tag{2.27}$$

where $Z' = Z_L/Z_0$. For any value of Z' a corresponding value of Γ_L is obtained, and in particular, when

$$Z_L = 0 \quad \text{then} \quad \Gamma_L = -1$$

$$|Z_L| = \infty \quad \text{then} \quad \Gamma_L = 1$$

$$Z_L = Z_0 \quad \text{then} \quad \Gamma_L = 0$$

According to the theory of complex functions, the mapping of Z' into Γ_L will be conformal, with the property that the angle between a pair of intersecting lines on the normalized impedance plane is maintained between the corresponding lines on the Γ plane. Lines of constant $\text{Re}(Z')$ and lines of constant $\text{Im}(Z')$ on the Z' plane map into families of circles on the Smith chart, with the two families intersecting at right angles. The Smith chart is sometimes referred to as the 'circle diagram'.

It follows from eqn. 2.27 that if $\text{Re}(Z') \geqslant 0$, then $|\Gamma_L| \leqslant 1$. Consequently, the right-hand half-plane in Fig. 2.10 is mapped onto or inside a circle of unit radius in the Γ plane.

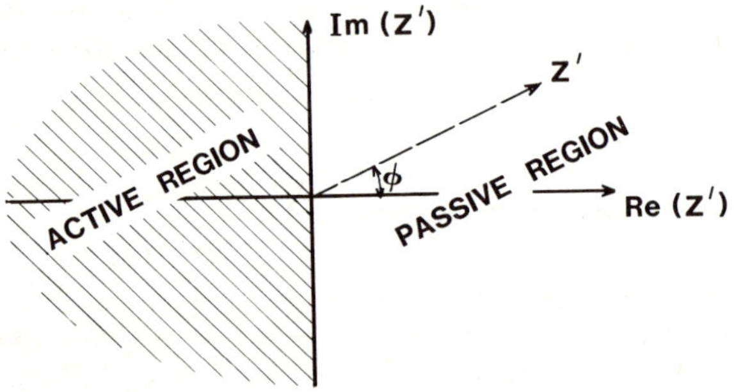

Fig. 2.10 *Complex impedance plane in Cartesian co-ordinates*

The Smith chart represents the complex Γ plane in Cartesian co-ordinates, and Γ_L may be read from it directly. The Cartesian grid system is not shown explicitly on a Smith chart – instead there is the system of circles which enable a particular value of Γ_L to be read off as the value of Z' related to it by eqn. 2.27. Conversely, a normalized impedance such as $Z' = 1 + j2$, shown in Fig. 2.11, can be plotted on the Smith chart using the circular co-ordinate system, and then the value $\Gamma_L = 0.5 + j0.5$ read directly from a Cartesian grid with origin at the centre, and scaled so that the Smith chart radius is unity.

2.5.3 *Special properties of the Smith chart*

Some of the properties of the Smith chart will become evident in other sections of this book. However two are worthy of special mention

(a) Variation of impedance along a transmission line.

The variation of impedance along a lossless transmission line is given by eqn. 1.15, and the corresponding reflection coefficient Γ_{in} is given by eqn. 1.19. Γ_{in} is the locus of a circle of radius $|\Gamma_L|$, with a centre which may be chosen to be the centre of the Smith chart. If the termination Z_L is passive and the normalization of the Smith chart is taken to be Z_0, the value of impedance at any point along the transmission line lies on this circle, and may be read directly from the chart co-ordinates. Progression around the circle in a clockwise direction from the point $Z' = Z_L/Z_0$

(representing the normalized termination) corresponds to movement towards the source, with one revolution representing a half (guide) wavelength movement along the transmission line.

Since, from eqn. 1.18 the locus of Γ_{in} for a uniformly lossy line is an exponentially shrinking spiral, a radial loss scale is often provided on Smith charts to allow the factor $\exp(-2\alpha l)$ to be taken into account.

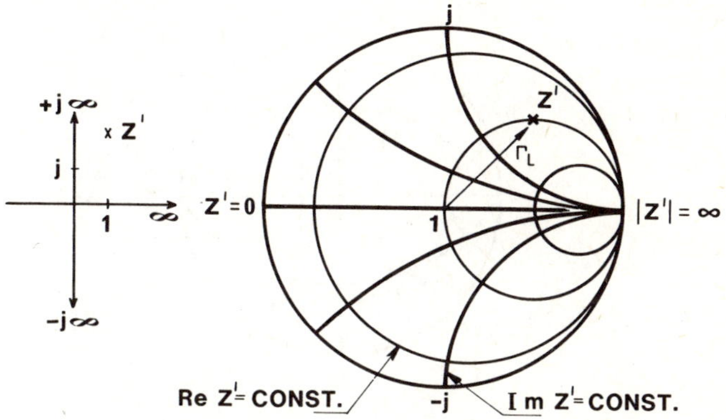

Fig. 2.11 *Smith chart or circle diagram, which transforms the passive impedance region of Fig. 2.10 into a circle*
The circular grid lines represent the locus of constant values of Re(Z') and Im(Z'). Reflection coefficients of impedances can be found by using a superimposed Cartesian Γ grid, which is not shown

(b) Inverse of a complex number.
The inverse of a complex number represented by a point on the Smith chart is the complex number represented by the mirror image of the point through the centre of the chart. This property follows from eqn. 2.27 after putting $Y' = 1/Z'$ to give

$$-\Gamma_L = \frac{Y'-1}{Y'+1} \tag{2.28}$$

The ease with which complex numbers may be inverted graphically using the Smith chart is useful for calculations where limited accuracy is sufficient. For example, the reduction of ladder networks may be performed quite rapidly by using the Smith chart to convert successively from admittance to impedance, and vice versa.

2.5.4 Carter chart

If lines of constant $|Z'|$ and lines of constant ϕ (ϕ being the argument of Z') are mapped from the Z' plane to the Γ plane using eqn. 2.27, the Carter chart shown in Fig. 2.12 is obtained [18]. Arcs of circles in the upper and lower half-planes with centres on the vertical axis represent lines of constant ϕ with the perimeter in the upper half-plane representing $\phi = \pi/2$, and the perimeter in the lower half-plane representing $\phi = -\pi/2$. Arcs of circles in the left and right half-planes with centres on the horizontal axis represent lines of constant $|Z'|$, with the exception of the perimeter circle.

Both the Smith chart and Carter chart have an implicit Cartesian grid system for reflection coefficient, but whereas the Smith chart co-ordinates allow Z' to be read directly in terms of its real and imaginary parts, the Carter chart co-ordinates allows Z' to be read directly in terms of its magnitude and phase.

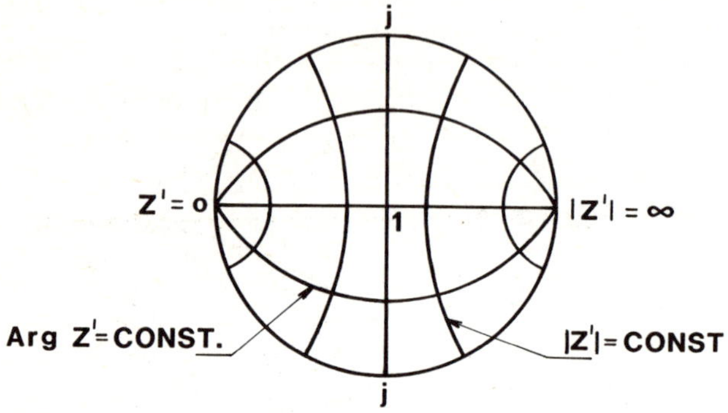

Fig. 2.12 *Carter chart, which transforms the passive impedance region of Fig. 2.10 into a circle*
The circular grid lines represent the locus of constant values of $|Z'|$ and arg (Z'). Reflection coefficients of impedances can be found by using a superimposed Cartesian Γ grid, which is not shown

References

1 KERNS, D.M. and BEATTY, R.W.: 'Basic theory of waveguide junctions and introductory microwave network analysis' (Pergamon Press, 1967)
2 DROPKIN, H.: 'Comments on "A rigorous technique for measuring the scattering matrix of a multiport device with a two-port network analyzer" ', and reply by TIPPET, J.C. and SPECIALE, R.A., *IEEE Trans.* 1983, **MTT-31**, pp. 79–81
3 TIPPETT, J.C. and SPECIALE, R.A.: 'A rigorous technique for measuring the scattering matrix of a multiport device with a 2-port network analyzer', *IEEE Trans.*, 1982, **MTT-30,** pp. 661–666
4 WOODS, D.: 'Multiport-network analysis by matrix renormalisation employing voltage-wave S-parameters with complex normalisation', *Proc. IEE,* 1977, **124,** pp. 83–86
5 WOODS, D.: 'Multiport network analysis by matrix renormalisation: extension to four ports', *Proc. IEE,* 1977, **124,** pp. 749–753
6 WOODS, D.: 'Multiport network analysis by matrix renormalisation: extension to 5- and 6-ports', *Proc. IEE,* 1978, **125,** p. 1217
7 BEATTY, R.W. and KERNS, D.M.: 'Relationships between different kinds of network parameters, not assuming reciprocity or equality of the waveguide or transmission line characteristic impedances', *Proc. IEEE,* 1964, **52,** pp. 84 and 426
8 GUILLEMIN, E.A.: 'Communication networks', Vol. II (John Wiley, 1935)
9 BOLINDER, E.F.: 'Note on the matrix representation of linear two-port networks', *IRE Trans.,* 1957, **CT-4,** pp. 337–339

10 LePAGE, W.R. and SEELY, S.: 'General network analysis' (McGraw-Hill, 1952)

11 KUHN, N,: 'Simplified signal flow graph analysis', *Microwave Journal,* 1963, **VI,** No. 11, pp. 59–66

12 WARNER, F.L.: 'Microwave attenuation measurement' (Peter Peregrinus, 1977), p. 276

13 MASON, S.J.: 'Feedback theory – some properties of signal flow graphs', *Proc. IRE,* 1953, **41,** pp. 1144–1156

14 MASON, S.J.: 'Feedback theory – further properties of signal flow graphs', *Proc. IRE,* 1956, **44**, pp. 920–926

15 MUNRO, N. and McMORRAN, P.D.: 'Signal-flow-graph reduction, Mason's rule and the system matrix', *Electron. Lett.,* 1970, **6,** pp. 752–754

16 SMITH, P.H.: 'A transmission line calculator', *Electronics,* 1939, **12,** pp. 29–31

17 SMITH, P.H.: 'An improved transmission line calculator', *Electronics,* 1944, **17,** p. 130

18 CARTER, P.S.: 'Charts for transmission-line measurements and computations', *R.C.A. Rev.,* 1939, **III,** pp. 355–368

Chapter 3

Basic circuit elements and components

In this chapter, some of the circuit components available for use in microwave impedance measuring networks are mentioned, and their properties discussed.

3.1 Connectors

The aim of the connector designer is to provide mechanical and electrical coupling between two sections of transmission line such that the electrical characteristics of the coupled line are indistinguishable from those of a continuous line. The principal electrical requirements are that the connector introduce the smallest possible reflections and losses, and design principles to achieve this have been formulated [1].

Practical connectors are never ideal, but they may still be used for precision impedance measurements provided that their electrical characteristics remain constant with repeated reconnection. If this is so, their electrical properties may be established, and used to account for the presence of the connector in subsequent measurements [2–5]. Apart from choosing a well-designed connector, the metrologist requiring high connector repeatability must ensure that the connector pair is clean, and joined with the correct mechanical loading. The tightening of a connector pair with the aid of a calibrated torque spanner is particularly important for small precision coaxial connectors, not only to achieve high repeatability, but also because some are easily damaged if overtightened.

The IEEE Standard for Precision Coaxial Connectors [6] includes a good set of references for methods of measuring coaxial connector properties.

At microwave frequencies, the most frequently encountered connectors are coaxial or waveguide, since these are the most common types of transmission line. Other transmission line types often use a transition to one of these connectors. Stripline circuits, for example, are often connected to external devices using a stripline-to-coaxial transition.

3.1.1 Coaxial connectors

Apart from connector reflection and loss, the additional parameters characterizing coaxial connectors are the maximum usable frequency (set by the cutoff frequency of the first propagating waveguide mode), leakage, electrical length, and d.c. contact resistance [6]. The basic design equation for coaxial line connectors is the one for the characteristic impedance

$$Z_0 = \frac{1}{2\pi}\sqrt{\left(\frac{\mu}{\epsilon}\right)}\ln\left(\frac{r_1}{r_2}\right) \tag{3.1}$$

where μ and ϵ are the permeability and permittivity of the medium, r_1 and r_2 are the respective outer and inner radii of the coaxial conductors as shown in Fig. 3.1*a*. For a reflectionless connection over a wide frequency band, Z_0 must be maintained as constant as possible at each transverse plane throughout the connector. This presents a difficult task to the designer because of the need for the inner conductor to be supported mechanically, therefore requiring the introduction of some dielectric material.* The effect of the presence of the dielectric on Z_0 may be compensated by a step change in the conductor diameters in the dielectric region, but this in turn introduces an equivalent shunt capacitance in the plane of the step. The magnitude of this 'step capacitance' has been calculated [7–10].

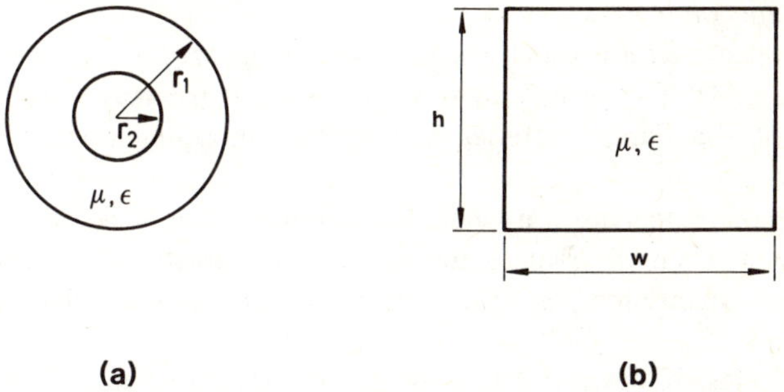

Fig. 3.1. *Cross-sectional geometries*
a coaxial transmission line
b rectangular waveguide

There are a large number of coaxial connector types in use [11–16], and many manufacturers produce the same nominal connector but with slightly different mechanical and electrical specifications. For example, the popular *N*-type connector has so evolved since its introduction that a recent study found 60 different makes and specifications to compare [17].

Connectors may be classified as to whether they are symmetrical (sexless) or asymmetrical (male/female, plug/jack). Of particular interest for the purpose of

* Notable exceptions are the types of line first introduced by the General Radio Co., with the inner conductor supported by the inner conductor of the connector to which it is mated. These 'beadless' lines are not suitable for general use, because two cannot be connected together.

impedance measurements is whether the connectors have a well-defined plane of mating. If they do not, the reference plane for the phase of the terminating impedance is ill-defined (see Table 3.1).

Table 3.1. *An indication of the improvement in connector performance over many years.*

Type	Approximate date of introduction	VSWR at 5 GHz (typical)	Approximate maximum usable frequency (GHz)	Symmetrical	Well defined plane of mating
GR 874	1948	1·035	7	yes	no
GR 900	1963	1·005	8.5	yes	yes
Prec. *N* type	1967	1·02	18	no	no
APC 7	1968	1·007	18	yes	yes
APC 3.5	1976	1·006	35	no	no

3.1.2 Waveguide connectors

Waveguide connections are usually made by bolting together flanges attached to the waveguide ends. The quality of the connection is partially determined by the conductivity of the flange material, and by the smoothness and flatness of the mating surfaces.

If the waveguide apertures are not dimensionally identical, or if they are not perfectly aligned when the flanges are mated, the characteristic impedance of the waveguide is not maintained and the connection will cause a reflected wave to be set up [18–23].

The magnitudes of the reflection coefficients set up when two waveguides are connected with a displacement s in either the E-plane (Fig. 3.2*a*), or the H-plane (Fig. 3.2*b*), were computed by Hunter [23], and are shown in Figs. 3.3 and 3.4, respectively. An approximate formula for these magnitudes is

$$\log_{10}(|\Gamma|) \simeq \sum_{n=0}^{3} u_n(\xi-\alpha)^n \log_{10}(\tau) + \sum_{n=0}^{3} v_n(\xi-\alpha)^n \tag{3.2}$$

where $\xi = h/\lambda_g$, $\alpha = 0{\cdot}3$, $\tau = s/h$ for E-plane displacement, and $\xi = w/\lambda$, $\alpha = 0{\cdot}7$, $\tau = s/w$ for H-plane displacement. The constants u_n and v_n are given in Table 3.2.

For E-plane displacements, eqn. 3.2 approximates $|\Gamma|$ with an error less than $1\frac{1}{2}\%$ for $0{\cdot}025 \leqslant s/h \leqslant 0{\cdot}25$, and with an absolute error of less than 0·0001 for $s/h < 0{\cdot}025$. For H-plane displacements, eqn. 3.2 approximates $|\Gamma|$ with an absolute error of less than 4% for $0{\cdot}6 \leqslant w/\lambda \leqslant 0{\cdot}9$, $0{\cdot}05 \leqslant s/w \leqslant 0{\cdot}25$. For $0{\cdot}6 \leqslant w/\lambda \leqslant 0{\cdot}95$, $s/w < 0{\cdot}05$, the approximation has an absolute error less than 0·0015, and for $0{\cdot}9 < w/\lambda \leqslant 0{\cdot}95$ the absolute error is less than 0·008 for $s/w \leqslant 0{\cdot}2$, and less than 0·015 for $s/w \leqslant 0{\cdot}25$.

The effects of E-plane or H-plane displacement are quite different in character.

E-plane displacements can be represented solely by a shunt susceptance in the displacement plane, but the effect of *H*-plane displacements requires both a shunt susceptance and an increase in electrical length of the waveguide for accurate representation.

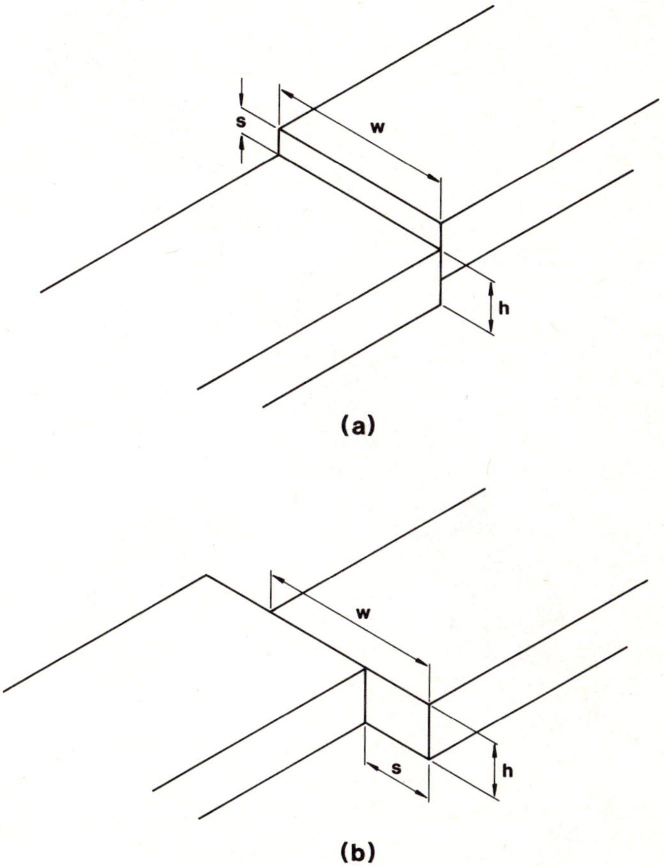

Fig. 3.2. *Rectangular waveguide displaced by an amount s in (a) the E-plane; (b) the H-plane*

3.1.3 Insertion loss and repeatability of connectors

Both the reflection coefficient and insertion loss of most microwave connectors are affected by the various forces acting on the contacting surfaces. Axial forces are required to hold the contacting surfaces together, but rotational and shear stresses should be avoided in order to produce predictable connector behaviour and prolong connector life. The connectors should be aligned without stress before being brought together. In the case of precision waveguide flange connections, special aligning holes and slightly tapered dowels are sometimes provided for preconnection alignment. For precision measurements, the contacting forces are standardized by the use of torque wrenches.

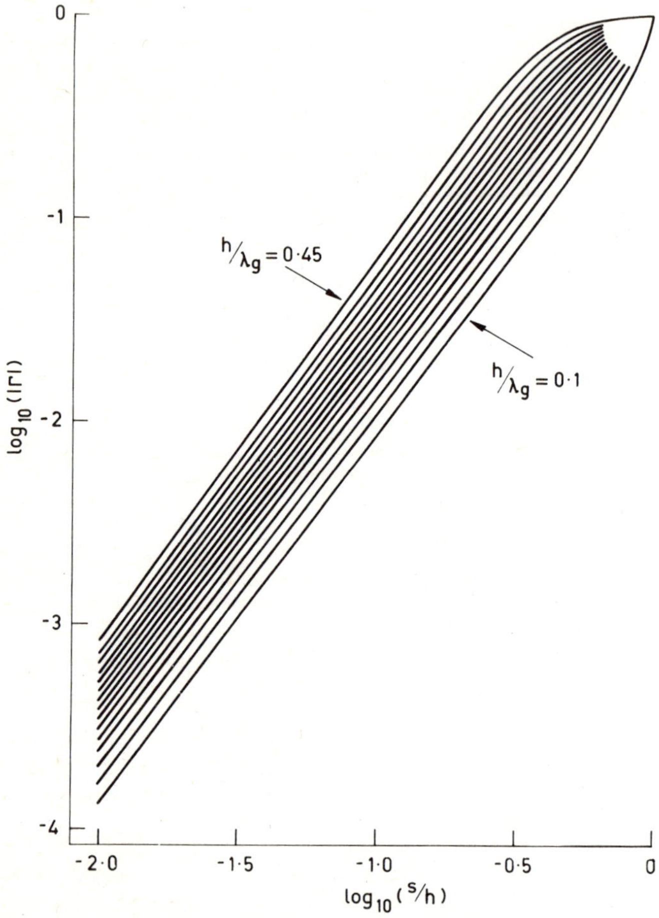

Fig. 3.3. ***Magnitude of the reflection coefficient caused by an E-plane displacement of s at a rectangular waveguide junction. The waveguide height is h***

[reproduced by permission of IEEE]

A means of facilitating rapid waveguide flange connections with standardized loading was reported by Johnson and Harvey [24], who used a hydraulically operated clamp.

Since the opening and closing of microwave connectors necessarily involves physical movement, some mechanical flexibility must be provided if a component has to be exchanged in an existing circuit. For precision measurements, it is unwise to use braided cables to provide this flexibility because of the instabilities they introduce. However, the so-called 'semi-rigid' cables (with flexible, solid copper conductors and PTFE dielectric) are well suited to this purpose, provided that sufficient lengths are used to keep the conductors within their elastic limit as they are flexed.

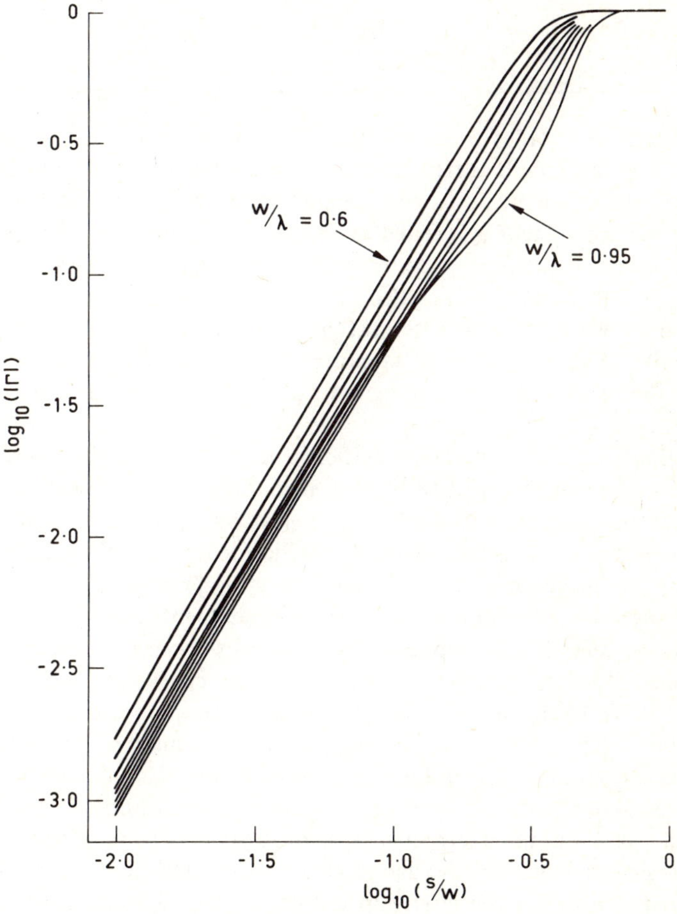

Fig. 3.4. *Magnitude of the reflection coefficient caused by an H-plane displacement of s at a rectangular waveguide junction. The waveguide width is w*

[reproduced by permission of IEEE]

A practical solution is to form a helical coil of semi-rigid cable, which will allow axial movement without causing fatigue.

Flexible waveguides formed into a 180° bend may be introduced into a

Table 3.2 *The constants used with eqn. 3.2 to compute the approximate reflection coefficient magnitude from displaced rectangular waveguides*

	E-plane displacement		H-plane displacement	
n	u_n	v_n	u_n	v_n
0	1·833	0·293	1·750	0·635
1	0·276	2·133	−0·332	−1·562
2	0·73	0·78	−2·71	0·44
3	0	19·69	−3·57	−7·63

waveguide circuit in order to permit the stress-free insertion of components. For example, a 1 m long length of flexible X-band waveguide bent with a 200 mm radius of curvature will allow most X-band components to be inserted into a microwave circuit. As the component is inserted or removed, the curvature and length of the curved section is maintained, thereby avoiding any significant signal amplitude or phase change with the movement [25].

Heavy microwave components must be supported to avoid connector stress. An alternative to rigid mechanical supports is to suspend the heavy component on springs, so that it 'floats' freely, and permits connections to be made with negligible stress.

Upper limits on both reflection coefficient magnitude and the insertion loss of microwave connectors are usually given in manufacturers' specifications. However, for precision measurements it is often the repeatabilities of these quantities which are the most important factors. A fixed insertion loss, and a fixed reflection coefficient may be taken into account, and the appropriate corrections made, but random changes are manifest as noise on the measurement results.

In order to quantify the repeatability of various connectors, the magnitude of the complex standard deviation of the reflection coefficient and the standard deviation of the insertion loss, were measured when connectors, which had had some years of laboratory use, were repeatedly disconnected and reconnected.* The results are given in Table 3.3. No special cleaning (or lapping for waveguide flanges) was given before the tests, and this may be the reason why the figures are, on average, a factor of ten worse than those of Pakay and Torok [26].

In practice, each waveguide flange connection introduces attenuation in the plane of the flange, and the reflection this causes can be represented in an equivalent circuit by a lumped series resistance [27, 28].

Somlo [28] used a tuned reflectometer to measure directly the insertion loss of waveguide flanges, by observing the return loss from a sliding, non-contacting short circuit which was moved across a flange connection between aluminium and copper waveguides. The variation of the return loss is shown in Fig. 3.5. The mean slopes on either side of the flange correspond to the guide attenuation constants,

*P. I. Somlo

Table 3.3 *Scatter of reflection coefficient magnitude and insertion loss with repeated connections of some 'used' connectors, measured at 8.2 GHz*

Connector type	$\lvert\sigma_{\lvert\Gamma\rvert}\rvert$	σ_α(db)	Torqued
GR 900	<0·001	<0·0005	yes
Type-N*	<0·06	<0·028	no
APC-7	<0·003	<0·01	yes
SMA	<0·03	<0·004	yes
X-band flange	<0·004	<0·0005	no

* non-precision, non-stainless steel

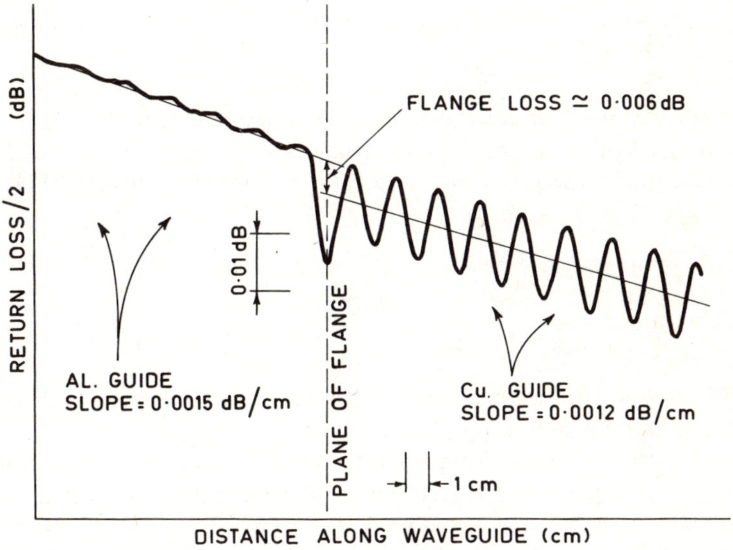

Fig. 3.5. *Direct measurement of waveguide flange loss*
The slopes indicate specific waveguide line losses, the offset of the mean slopes is the lumped flange loss. The ripple is caused by the lumped loss, as evidenced by the phasing of the pattern
[reproduced by permission of IEEE]

and show directly the difference between aluminium and copper waveguides. The offset of the sloping lines at the flange is due to the lumped (series) resistance in the plane of the flange coupling, and is caused by the imperfect electrical connection at the flanges. The increased ripple when the sliding short circuit has passed over the flange connection away from the source is due to the reflection caused by the series lumped loss. This is shown by the phasing of the ripple, which indicates maximum loss each time the short circuit is an integral number of half guide wavelengths away from the flange plane. (If the increased ripple had been due to a reactive type of mismatch, such as caused by a misaligned flange connection [23], the ripple pattern would be shifted by 90° from that measured.)

3.2 Precision transmission line sections

A section of uniform transmission line supporting a particular propagating mode is fully described by its length, propagation constant, characteristic impedance, and the cutoff frequency of that mode. If such a section is made to precise mechanical specifications, and the losses are kept as low as possible by a judicious choice of dielectric and conductor materials, the characteristic impedance and the phase shift introduced by the section can be calculated, to a first order of accuracy, from dimensional measurements. Higher accuracy is obtained by taking the losses into account. Precision line sections may therefore be used as characteristic impedance standards, phase shift standards, or attenuation standards in the case of a waveguide below cutoff [29, 30].

3.2.1 Use of precision transmission lines as zero reflection standards

The use of a length of precision coaxial line as an impedance standard, or of a length of precision waveguide as a zero reflection standard, is based [31–34] on eqns. 1.16 and 1.18

$$\begin{aligned}\Gamma_{in} &= \Gamma_L \exp(-2\gamma l)\\ &= \frac{Z_{in} - Z_0}{Z_{in} + Z_0}\end{aligned} \tag{3.3}$$

If the precision line is terminated in its characteristic impedance, then $\Gamma_L = 0$, and for any line length $\Gamma_{in} = 0, Z_{in} = Z_0$.

In practice it is difficult to construct a termination with an impedance identical to Z_0, but it is rather easier to devise a termination with an impedance near Z_0. If such a termination is constructed so that it can be slid inside the precision line section, it follows from eqn. 3.3 that for a loss-free line, the locus of Γ_{in} is a small circle on the complex Γ plane (Smith chart). The centre of this circle approximates closely the location of $\Gamma_{in} = 0$, but any connector reflection will cause the centre to be offset from the origin. 'Beadless' coaxial sliding terminations, where the inner conductor is supported by the connector to which it is mated, are an attempt to minimize connector reflection.

At microwave frequencies, low reflection, wideband, jitter free, repeatable waveguide sliding terminations are realized using lossy dielectric material (wood, epoxy resin loaded with iron powder, silicon carbide) which is tapered physically, and installed so as not to contact the walls of the precision line. Alternatively, the sliding termination can be fabricated from tapered cards having resistance per square of 200–300 ohms.

Any variations in dimensions along the length of precision transmission line will cause the locus of Γ_{in} on the complex Γ plane to be non-circular as the termination is slid along. In such cases, the termination should be slid over a distance of several half-wavelengths so that a better estimate of the central position corresponding to $\Gamma_{in} = 0$ is obtained. The locus of Γ_{in} using a typical commercial coaxial sliding load is shown on the complex Γ plane in Fig. 3.6.

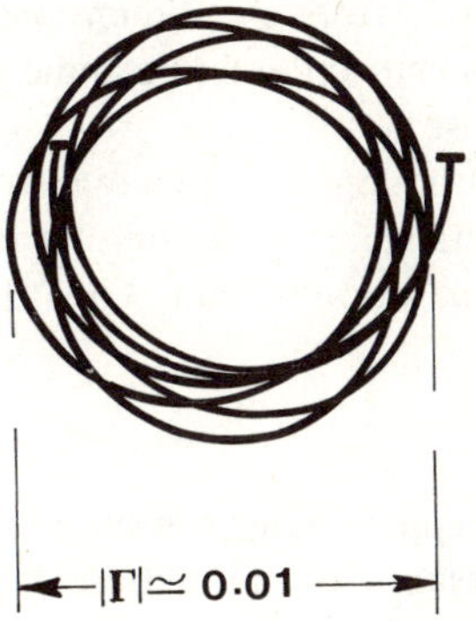

Fig. 3.6. *Measured variation of the reflection coefficient* Γ *of a commercial 50 ohm coaxial sliding load*

At frequencies below about 1 GHz, the use of coaxial sliding terminations becomes impractical because of the increased wavelength. However, if a $\lambda/4$ section of precision line is available, it can be inserted in front of a nominally matched terminating resistor, and the two points obtained on a Smith chart display averaged. At even lower frequencies, where $\lambda/4$ lines are impractical, the d.c. resistance of terminations designed to be wideband can be used. Confidence in the design of the termination can be obtained by checking the d.c. resistance value against that measured at frequencies where $\lambda/4$ lines are available.

By using precision coaxial line in conjunction with a two-port resistor and a termination, both the magnitude and the phase of the impedance at the input port may be varied [35]. This technique is a means of providing a range of impedance standards.

3.2.2 Use of precision transmission lines as non-zero reflection standards

The ratio of the characteristic admittances of two coaxial lines of different dimensions but filled with the same dielectric material is, from eqn. 3.1, for TEM mode propagation

$$Y' = \frac{Y_1}{Y_0} = \frac{\ln(r_1/r_2)}{\ln(r_1'/r_2')} \tag{3.4}$$

If two such lines are connected, and the second is terminated in a match or sliding load, the reflection coefficient in the connection plane, as seen from the unterminated line, is

$$\Gamma = \frac{1 - Y' - jB'}{1 + Y' + jB'} \tag{3.5}$$

where B' is the susceptance (normalized to Y_0) corresponding to the step capacitance introduced by the abrupt dimensional changes at the connection [7–10].

Because eqn. 3.5 is calculable from the dimensions of the coaxial lines, it may be used as the basis for a non-zero reflection standard. In practice, it is convenient for ease of connection to choose $r_1 = r_1'$.

The reflection from the junction of a rectangular waveguide of width w and height h (Fig. 3.1b) and a rectangular waveguide having the same width but height h', is given for TE_{10} mode propagation by eqn. 3.5 with

$$Y' = \frac{h}{h'} \tag{3.6}$$

and B' being the susceptance appropriate to waveguide steps [36]. In practice, it is usual to choose $h' < h$ and realize the reflection standard from a reduced height waveguide.

The use of a reduced height section of precision waveguide terminated in a matched length of normal height waveguide has been advocated as a reflection standard [37]. If the reduced height section is an odd multiple of a quarter guide-wavelength long, the effects of the step susceptances tend to cancel.

The use of two lengths of precision waveguide offset at a flange junction has also been proposed as a reflection standard [23]. Such a junction is easy to construct, and can be made adjustable. It is possible to describe the effect of wall deformation on the impedance of a length of precision waveguide sufficiently well for it to be used as a reflection standard [38].

3.2.3 Use of precision transmission lines as phase-shift standards

In principle, any section of uniform transmission line may be used as a phase-shift standard, provided that the phase constant and the length of the section are well defined. Precision fixed length sections are used as phase-shift standards in both coaxial and waveguide systems, but the effects of connector reflection and poor repeatability can contribute significantly to the measurement uncertainty.

Alternatively, adjustable phase-shift standards may be constructed by varying the effective length of the section. This is done most conveniently by sliding a short circuit (preferably non-contacting) in a precision section of line, and measuring the position with a high resolution micrometer or laser facility [39]. In order to fully realize the potential accuracy of a precision section, residual reflections must be tuned out.

The propagation constant in hollow metallic pipes of rectangular or circular cross-section is given by Kerns and Hedberg [40] and Schaffeld and Bayer [41]. Fig. 3.7 depicts the dependence with frequency of the attenuation constant α, and the phase-shift constant β, for a cylindrical copper pipe (designed to be used as a standard attenuator at 30 MHz, where α is virtually constant).*

3.2.4 Factors affecting the accuracy of precision line standards

The reflection coefficient obtained using precision transmission lines as standards is affected by various factors including connector loss, connector misalignment, and the finite conductivity of the line. The departure of the precision line from the ideal geometry, including conductor ellipticity and sagging of the centre conductor, has an effect on the characteristic impedance [5].

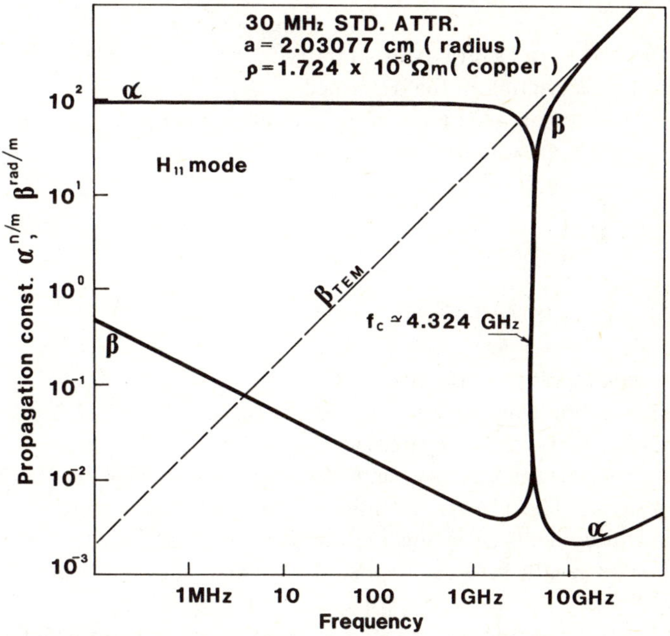

Fig. 3.7. *Propagation constant of a cylindrical copper pipe of radius a and resistivity ρ used as a 30 MHz standard attenuator*

The use of beadless connectors has already been mentioned as a means of reducing the reflection from a coaxial junction. The loss associated with a waveguide flange, or a coaxial connector with a well-defined mating plane, can be represented in an equivalent circuit as a series resistor at the connector plane, and it therefore causes a direct error in the calculated reflection of a precision line standard. In

* It is interesting to note that although for a perfectly conducting pipe $\beta = 0$ at frequencies below cutoff, when the pipe has finite wall conductivity and therefore wall losses, $d\beta/df < 0$ below cutoff. This indicates that in a practical waveguide below cutoff the evanescent mode is a heavily attenuated backward wave, where the phase and group velocities have opposite sign [42].

order to minimize the effect of flange loss for reduced height waveguide standards, the step in height should occur an odd multiple of one-quarter of a guide wavelength from the flange, so that the reflected wave caused by flange loss is in phase quadrature with the reflected wave due to the change in waveguide height [28].

Most geometrical discontinuities related to the connection of two waveguides can be treated by numerical methods [43, 44]. For waveguides supporting TE_{10} mode propagation, the mean width of the guide should be taken as representative if there is a small variation in width with height. For a small variation of height with width, the effective height h_e is related to the nominal height h by

$$h_e/h = \int E^2 dS' \Big/ \int E^2 dS \tag{3.7}$$

where E is the transverse electric field intensity in the nominal guide of cross-section S, and S' is the cross-section of the perturbed guide [36].

The reflection set up at the junction of a rectangular waveguide and one with fillets of radius r is approximated by [38]

$$\Gamma = \left[\frac{\lambda_g r}{w}\right]^2 \frac{(4-\pi)}{8wh} \tag{3.8}$$

where λ_g is the guide wavelength.

3.2.5 Finite conductivity – skin effect

If a transmission line conductor is lossless, the (external) E field is normal to the surface, and no energy is dissipated in the conductor. However, practical conductors are of finite conductivity, and dissipate energy which is thus lost to the propagating signal. The effects of finite conductivity on a parallel plate transmission line will be deduced in this section using an approximate theory given by Howe [45]. The results will be applied to coaxial lines and waveguides in the following sections.

The parallel plate waveguide in Fig. 3.8 has metal conductors of infinite extent, and supports a plane (TEM) wave travelling from left to right, with the E and H fields directed as shown. Consider a unit square on the surface of one of conductors, with two edges normal to the direction of propagation. The current in the metal flows almost parallel to the surface and in the direction of propagation, and since the conductivity is finite, there is a potential drop between the two edges normal to it. The E and H fields in the metal are almost parallel to the surface, and a plane wave propagates into the metal.

We may imagine this plane wave to be supported by a parallel plate transmission line with perfect conductors A and B penetrating into the metal. The impedance measured between the upper edges of A and B is termed the surface impedance, Z_s, and is calculable as the input impedance of the imagined transmission line. Since the plates of this transmission line are perfectly conducting, the resistance per unit length (depth into the metal) is zero. The inductance per unit length is $\mu_m = \mu_c \mu_0$, where μ_c is the relative permeability of the conductor, and the conductance per unit length is equal to σ_m, the conductivity of the conductor. The capacitance per unit length is $\epsilon_m = \epsilon_c\epsilon_0 = \epsilon_c/(c^2\mu_0)$, where ϵ_c is the relative permittivity of the conductor. Since, for good metal conductors at microwave frequencies, $\sigma_m \gg \omega\epsilon_m$, the expressions for characteristic impedance Z_m, and propagation constant γ_m of this imagined transmission line reduce, from eqn. 1.6, to

$$Z_m \simeq \left(\frac{j\omega\mu_m}{\sigma_m}\right)^{1/2} = (1+j)\left(\frac{\omega\mu_m}{2\sigma_m}\right)^{1/2}$$

$$\gamma_m \simeq (j\omega\mu_m\sigma_m)^{1/2} = (1+j)\left(\frac{\omega\mu_m\sigma_m}{2}\right)^{1/2} \tag{3.9}$$

A similar parallel plate line can be imagined next to the first, sharing a common plate B. The two currents flowing in this plate oppose and cancel, so we can now imagine removing the plates without causing any effect. There is therefore no resultant current normal to the conductor surface, in close approximation to the true behaviour.

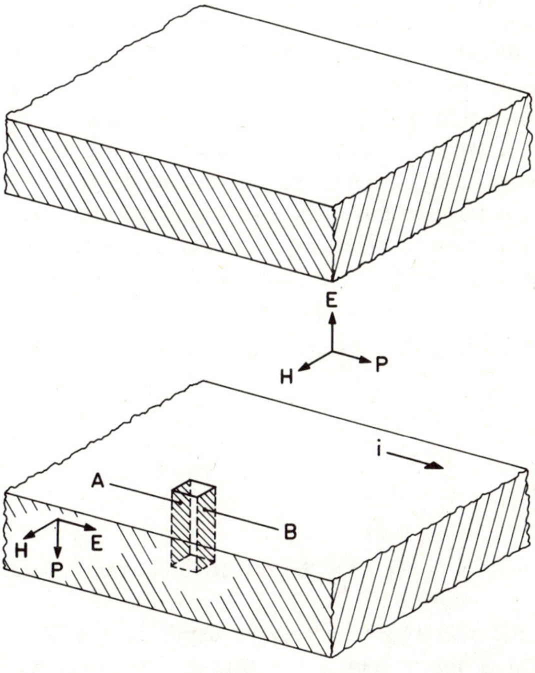

Fig. 3.8. *Section of an infinite parallel plate transmission line showing the construction used to derive the surface impedance of the conductor. A and B are unit width surfaces of an imagined parallel plate transmission line inside the metal.*

The propagation constant γ_m has equal real and imaginary parts for plane wave propagation into the conductor, and is large in comparison with those of practical transmission lines. The wave is therefore highly attenuated, and the velocity of propagation ω/β_m is many orders of magnitude smaller than in vacuum. The real part of γ_m is α_m, the attenuation constant in nepers/metre, and its inverse in metres/neper, is termed the skin depth, δ. Therefore one skin depth is the distance a wave penetrates into the conductor while being attenuated one neper (a factor of e = 2·718), and is given by

$$\delta = \left(\frac{2}{\omega\mu_m\sigma_m}\right)^{1/2} \tag{3.10}$$

Since the imaginary part of the propagation constant has the same value as the real part, the wavelength in the conductor is $2\pi\delta$. Because this wavelength is extremely small, the curvature of cylindrical conductors in all but the electrically thinnest wires may be disregarded, and propagation through metal conductors can be treated as planar. If δ is comparable with or greater than the radii of the conductors, the expression for Z_m in eqn. 3.9 is not valid, and cylindrical wave propagation through the metal must be considered [46].

Propagation into the metal is similar to propagation along a lossy transmission line, and consequently, if the metal is more than a few skin depths thick (return loss greater than 40 dB), the surface impedance Z_s closely approaches Z_m, the characteristic impedance of the metal.

It should be noted that the surface impedance of a lossy conductor always has an inductive component, which may be regarded as additional to that of a similar transmission line formed by perfectly conducting metal. In order to combine these inductances, we can imagine the surface of the metal recessed below the actual surface, and the lossy metal replaced by a perfect conductor. The recession depth Δr is that thickness of material filling the line which has the same reactance as Z_m,

$$\omega\mu\Delta r = \left(\frac{\omega\mu_m}{2\sigma_m}\right)^{1/2} \tag{3.11}$$

which leads to

$$\Delta r = \frac{\delta\mu_m}{2\mu} \tag{3.12}$$

As far as inductance is concerned, the effective electrical surface is therefore a distance Δr below the physical surface.

For non-magnetic metals, the recession depth is one half the skin depth, provided the metal is more than a few skin depths thick. As the frequency is reduced, the recession depth for metals of finite thickness t approaches $t/3$, and that for cylindrical conductors of radius r approaches $0{\cdot}221r$ [47].

In skin effect calculations it is usually assumed that: (*a*) the surfaces are smooth, and (*b*) the conductivity of the metal is known. If the surface roughness (expressed as deviation from the mean) is comparable with the skin depth, the surface currents have longer paths than on a smooth surface, and this increases losses. Electroplated surfaces are usually less dense than the bulk metal, and without special plating techniques the bulk conductivity is rarely realized. In practice, therefore, the measured loss may be (typically) 30% higher than that computed from idealized skin-effect theory.

3.2.5.1 Effect of finite conductivity on coaxial lines: It is convenient to express the characteristic impedance Z_0 and propagation constant γ of a practical coaxial line, made from metal of finite conductivity, in terms of the characteristic impedance Z_0^∞ and propagation constant γ^∞ of a line having the same geometry, but made from metal of infinite conductivity. It follows from eqn. 1.6 that

$$Z_0^\infty = \left[\frac{j\omega L}{G + j\omega C}\right]^{1/2} \tag{3.13}$$

$$\gamma^\infty = [j\omega L(G + j\omega C)]^{1/2}$$

It can be seen from Fig. 3.8 that the surface impedance Z_s associated with practical conductors impedes the axial current flow in the main transmission line, and therefore may be regarded as a series impedance

$$Z_r = R_r + jX_r \tag{3.14}$$

per unit length of the line [48, 49], enabling Z_0 and γ to be written as

$$Z_0 = \left[\frac{R_r + j(X_r + \omega L)}{G + j\omega C}\right]^{1/2} = Z_0^\infty \left[1 + \frac{Z_r}{j\omega L}\right]^{1/2}$$

$$\gamma = [\{R_r + j(X_r + \omega L)\}(G + j\omega C)]^{1/2} = \gamma_0^\infty \left[1 + \frac{Z_r}{j\omega L}\right]^{1/2} \tag{3.15}$$

From the geometry of the coaxial line in Fig. 3.1*a*,

$$Z_r = \frac{1}{2\pi}\left(\frac{Z_{s1}}{r_1} + \frac{Z_{s2}}{r_2}\right) \tag{3.16}$$

where Z_{s1} and Z_{s2} are the surface impedances of the outer and inner conductors, respectively.

For precision air-dielectric coaxial lines, $G = 0$, and therefore

$$Z_0^\infty = \left(\frac{L}{C}\right)^{1/2} \tag{3.17}$$

which is also the limiting value of the characteristic impedance of a lossy line as the frequency becomes infinite, and is dependent solely on the line geometry. Then, from eqn. 3.15, using eqns. 3.9 and 3.16

$$Z_0 \simeq Z_0^{\infty} + \frac{(1-\mathrm{j})\omega}{2\pi k}\left(\frac{\delta_1\mu_{m1}}{r_1} + \frac{\delta_2\mu_{m2}}{r_2}\right) \tag{3.18}$$

and the attenuation and phase constants of the line are

$$\begin{aligned} \alpha &\simeq \frac{k}{4\mu \ln (r_1/r_2)}\left(\frac{\delta_1\mu_{m1}}{r_1} + \frac{\delta_2\mu_{m2}}{r_2}\right) \\ \beta &\simeq k + \alpha \end{aligned} \tag{3.19}$$

provided $\alpha \ll k$, where $k = \beta^{\infty} = \omega(\mu\epsilon)^{1/2}$. The subscripts 1 and 2 refer to the outer and inner conductors, respectively. These results are available in graphical form for the four sizes of air-dielectric coaxial transmission line commonly referred to as 7 mm, 14 mm, 21 mm, and 3/4 inch [50].

If the conductors of the coaxial line have been electroplated with successive layers of different metals (perhaps silver to increase the conductivity, and gold to protect the silver), the surface impedance Z_s is calculable by treating the conductors as multisectioned transmission lines where propagation into the metal is considered [51, 46]. Provided the bottom layer is more than several skin-depths thick, its impedance Z_1 can be calculated from eqn. 3.9 and regarded as the termination of the next metallic layer of impedance Z_2 and propagation constant $\gamma_2 = (\mathrm{j}\omega\mu_2\sigma_2)^{1/2}$. The thickness of this second layer may be used in the calculation of the impedance at its surface. After repeating the process for each subsequent layer, Z_s may be calculated. If the thickness of the bottom layer is comparable with or less than several skin depths, a terminating impedance of 377 ohms should be used when the layer is air backed.

The results of such a computation are shown in Fig. 3.9 for a 3 mm brass base onto which a 0·0127 mm (0·0005 in) thick silver layer is electroplated, and topped with a 0·00127 mm (0·00005 in) layer of gold. Plane wave propagation is assumed, and bulk conductivity figures are used. Fig. 3.9 shows how the (air backed) brass affects the surface impedance at the lower frequencies, and how it is only at the higher frequencies that the composite conductor behaves as a single metal, with real and imaginary parts of Z_s equal, in accordance with eqn. 3.9. At these higher frequencies it can be concluded that the current is effectively confined to the gold layer.

Components of the skin-effect correction are plotted in Fig. 3.10 for a nominal 50 ohm coaxial line, with conductor diameters of $2r_1 = 14{\cdot}288$ mm and $2r_2 = 6{\cdot}204$ mm, where both conductors are brought to these dimensions by electroplating 0·0254 mm (0·001 in) silver and then 0·000254 mm (0·00001 in) gold onto brass stock. Above 10 MHz, the presence of the brass base in the electroplated conductor has an insignificant effect, and the current is almost completely confined to the silver and gold layers.

The factor X_r/ω can be regarded as a modification to the lossless line inductance L. The recession depths Δr_1 and Δr_2 for the outer and inner conductors are calculable by equating the extra reactance introduced by the recessing procedure to that due to the losses [47]. Thus

$$\begin{aligned} \frac{\omega\mu}{2\pi}\left[\ln\left(\frac{r_1+\Delta r_1}{r_2}\right)-\ln\left(\frac{r_1}{r_2}\right)\right] &= \frac{X_{m1}}{2\pi r_1} \\ \frac{\omega\mu}{2\pi}\left[\ln\left(\frac{r_1}{r_2-\Delta r_2}\right)-\ln\left(\frac{r_1}{r_2}\right)\right] &= \frac{X_{m2}}{2\pi r_2} \end{aligned} \qquad (3.20)$$

which with eqns. 3.9 and 3.10 give

$$\begin{aligned} \Delta r_1 &= r_1\left[\exp\left(\frac{\delta_1\mu_{m1}}{2r_1\mu}\right)-1\right] \\ \Delta r_2 &= r_2\left[1-\exp\left(\frac{-\delta_2\mu_{m2}}{2r_2\mu}\right)\right] \end{aligned} \qquad (3.21)$$

As r_1 or r_2 becomes large, these expressions reduce to that in eqn. 3.12 for the recession depth of a plane conductor.

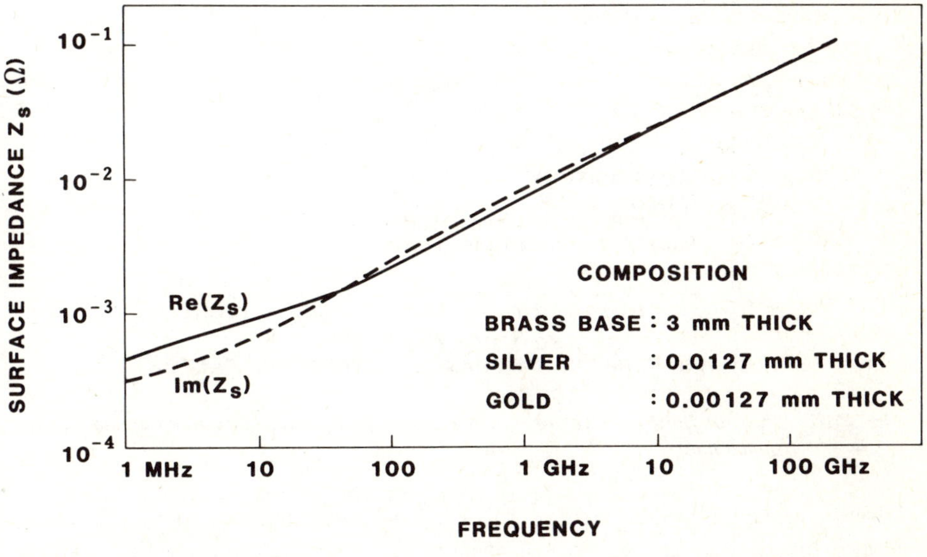

Fig. 3.9. *Surface impedance of a 3 mm brass plate onto which have been electroplated silver and gold layers*

3.2.5.2 Effect of finite conductivity on waveguides: The characteristic admittance Y_0 for TE_{10} mode propagation in the rectangular waveguide of Fig. 3.1*b* filled with material of permeability μ and permittivity ϵ is [38]

$$Y_0 = Y_0^\infty + \Delta Y_0^\infty$$

$$\Delta Y_0^\infty = Y_0^\infty(\mathrm{j}-1)\frac{\delta\mu_m}{\mu}\left[\frac{1}{2h} - \left(\frac{\pi}{\beta^\infty w}\right)^2\left(\frac{1}{w}+\frac{1}{2h}\right)\right] \tag{3.22}$$

where the skin depth δ is given by eqn. 3.10 with σ_m being the conductivity and μ_m the permeability of the walls. Y_0^∞ is the characteristic admittance of a dimensionally identical waveguide with lossless walls, and

$$(\beta^\infty)^2 = k^2 - \left(\frac{\pi}{w}\right)^2, \quad k^2 = \omega^2\mu\epsilon \tag{3.23}$$

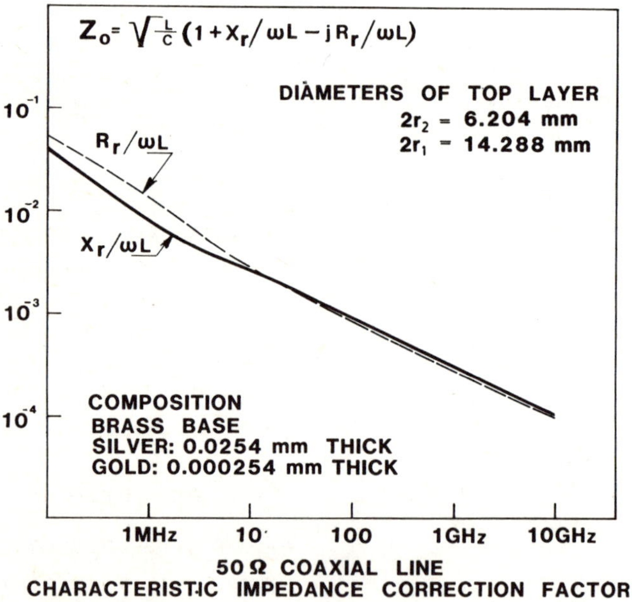

Fig. 3.10. *Effect of finite conductivity on the characteristic impedance of a nominal 50 ohm coaxial line*

If two waveguides having the same cross-sectional dimensions but with walls characterized by σ_1, μ_{m1} and σ_2, μ_{m2}, respectively, are connected, the reflection coefficient in the first waveguide caused by the change in wall properties at the connection is, from eqns. 1.16 and 3.22,

$$\Gamma = (\mathrm{j}-1)\,\frac{(\delta_1\mu_{m1}-\delta_2\mu_{m2})}{2\mu}\left[\frac{1}{2h}-\left(\frac{\pi}{\beta^\infty w}\right)^2\left(\frac{1}{w}+\frac{1}{2h}\right)\right] \tag{3.24}$$

where use has been made of the fact that $\Delta Y_0^\infty \ll Y_0^\infty$ for typical waveguides. The propagation constant for the lossy-walled waveguide is given by [40, 52]

$$\gamma = \alpha + \mathrm{j}\beta = \left[(\gamma^\infty)^2 - (1-\mathrm{j})\,\frac{\delta\mu_m}{\mu}\left(\frac{2\pi^2}{w^3}+\frac{k^2}{h}\right)\right]^{1/2} \tag{3.25}$$

where $\gamma^\infty = \mathrm{j}\beta^\infty$ is the propagation constant in the equivalent lossless guide. Above the cutoff frequency of the waveguide

$$\begin{aligned} \alpha &\simeq \frac{\delta\mu_m}{2\beta^\infty\mu}\left(\frac{2\pi^2}{w^3}+\frac{k^2}{h}\right) \\ \beta &\simeq \beta^\infty + \alpha \end{aligned} \tag{3.26}$$

provided $|\alpha| \ll |\beta^\infty|$, valid for practical waveguides.

3.3 Transmission line junctions

The general theory of transmission line junctions has been dealt with extensively by Kerns and Beatty [53] and is outside the scope of this book. However, in this section the properties of transmission line junctions will be discussed in so far as they relate to impedance measurements.

Transmission line junctions may be classified according to the number of ports (connectors) they possess. A linear N-port is characterized by N^2 S parameters associated with N characteristic port impedances, as discussed in Chapter 2. A linear reciprocal N-port is characterized by $\frac{N}{2}(N+1)$ S parameters if they are all associated with the same characteristic port impedance.

3.3.1 One-ports

A one-port is a microwave network element which terminates a transmission line, and is designed to create a predictable reflection coefficient. One-ports are, most frequently, intended to act as absorbing loads ($\Gamma = 0$), and dissipate all power incident upon them. Such dummy loads are used in the construction of other microwave circuit components, including three-port directional couplers in which the fourth port is internally terminated, and isolators made from circulators. Other one-ports are designed to be fully reflective ($|\Gamma| = 1$), and in this role are used in the construction of some resonators and filters. Fully absorbing or reflecting one-ports are also useful as standards for calibrating reflectometers. Some one-ports are designed to have a predictable intermediate value or reflection coefficient ($0 < |\Gamma| < 1$) and are used as reflectometer standards.

3.3.1.1 Matched loads: Fully absorbing one-ports are normally referred to as matched loads, and are realized as either discrete resistors (chips or discs), or in distributed form as suitably tapered resistance cards or films, or lossy dielectric material. Discrete (lumped) resistors are not easily realized in waveguide, and therefore only the distributed forms are used.

Matched loads are characterized by the type of transmission line and connector for which they are designed, and their power handling capability. Nominally matched loads (those for which $|\Gamma|$ is close to but not necessarily zero) are characterized additionally by $|\Gamma|$ (or VSWR) as a function of frequency. As already discussed in Section 3.2.1, a perfectly matched load is difficult to construct, but the effect of connecting a perfect match can be estimated by sliding a nominally matched load inside a length of precision transmission line. The average of the impedance over a half-wavelength movement approximates closely the characteristic impedance of the precision line.

3.3.1.2 Fully reflecting one-ports: Any impedance lying on the outer perimeter of the Smith chart ($\mathrm{Re}(Z) = 0$) is fully reflecting. At microwave frequencies such impedances may be realized approximately by terminating low-loss transmission lines in short or open circuits.

Short circuits in coaxial line or waveguide can be either fixed in position or movable. Transmission lines having connectors with a well-defined mating plane (for example, GR-900, APC-7, plane waveguide flanges) are easily terminated in fixed short circuits by connecting a flat metal plate of high conductivity across them. Other connector types (for example, GR-874, *N*-type), for which the effective electrical plane of the short circuit is ill-defined, are not recommended for precision impedance measurements.

The effectiveness of a short circuit (how nearly $|\Gamma|$ approaches unity) depends primarily on the contact loss, the loss due to the finite bulk conductivity of the metal of the short circuit, and also the connector loss, as discussed in Section 3.1.3. Contact loss can be minimized by choosing metal of high conductivity for the contacting surfaces, and ensuring that it has a high quality finish which is kept clean.

The effect of finite bulk conductivity of the short circuit can be dealt with theoretically. The reflection coefficient in the plane of the short circuit is given by

$$\Gamma = \frac{Z_s - Z_w}{Z_s + Z_w} \tag{3.27}$$

where Z_s is the surface impedance of the metallic short circuit (Section 3.2.5). Z_w is the wave impedance, which is $\eta = (\mu/\epsilon)^{1/2}$ for TEM propagation, ($\simeq 377$ ohms in air) and for waveguide mode propagation is, from eqns. 1.42 and 1.43,

$$Z_w^{TM} = \eta \left[1 - \left(\frac{f_c}{f}\right)^2\right]^{1/2} \tag{3.28}$$

$$Z_w^{TE} = \frac{\eta}{\left[1 - \left(\frac{f_c}{f}\right)^2\right]^{1/2}} \tag{3.28}$$

As an illustration of the use of eqn. 3.27, the magnitude of the reflection coefficient of a coaxial line flat metallic short circuit was computed as a function of frequency, taking the short circuit as having a brass base, electroplated with silver to 0·01 mm thickness, with a final gold flash of 0·001 mm thickness. The results, shown in Fig. 3.11, indicate that the effectiveness of the short circuit falls with frequency (due to increasing skin-effect losses), but at 10 GHz, $|\Gamma| \simeq 0{\cdot}9998$, which is still close enough to unity and to be taken as such for all but the most demanding accuracy requirements. Because the surface impedance has an inductive component, the effective plane of the short circuit is recessed by half a skin depth, as shown in Section 3.2.5. At 10 GHz, the recession depth for the above example is 0·0004 mm, which is negligible for most practical cases, since it is much less than the wavelength in the coaxial line (less than 0·01 electrical degrees).

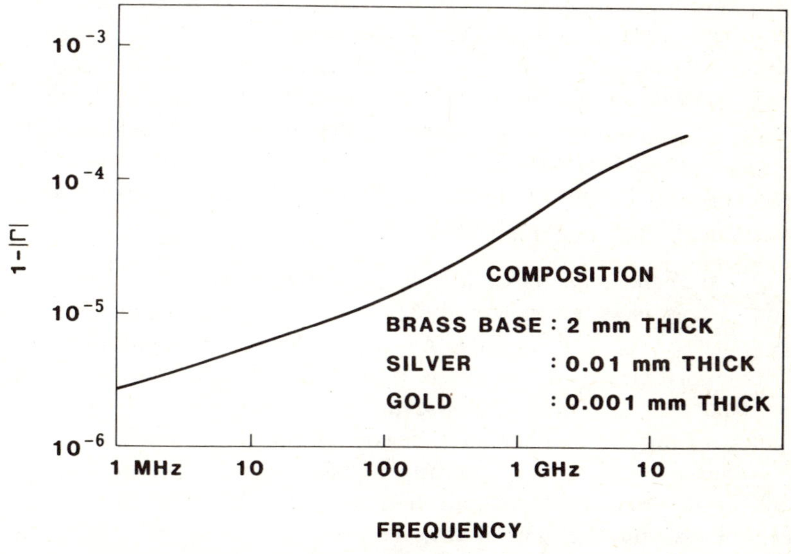

Fig. 3.11. *Computed reflection coefficient of a plane coaxial short circuit of electroplated composition, as a function of frequency. Contact loss is not included*

Sliding short circuits are often used in impedance measurements. They are constructed from precision sections of transmission line, with a slideable element designed to cause as full a reflection as possible. In coaxial line, sprung metal fingers connected to each other by a metallic annulus are used to contact both conductors. Contacting fingers are also used in waveguides, but non-contacting choke designs offer greater repeatability. The losses of sliding short circuits have two components in addition to connector loss (Section 3.1.3). A constant loss associated with the sliding element comprises skin loss and (if applicable) contact loss. A variable loss component depends on the length of transmission line between the short-circuit element and the connector port. For precision measurements, coaxial sliding short circuits can be made using 'beadless' connectors, with the inner conductor supported at the connector port only by the inner conductor of the mating connector. This design eliminates most connector discontinuities, and an uninterrupted, uniform transmission line is presented at the connector plane. It is important that an absorbing element be located in the transmission line behind the sliding short-circuit element in order to dampen any resonances that may otherwise occur. Because power can leak past the sliding element into the cavity beyond, it is possible that, without adequate damping, this cavity will resonate and have a large effect on the performance of the sliding short circuit.*

Open circuits are more difficult to realize than short circuits, first because open lines tend to radiate, and secondly because the effective position of the open circuit is more difficult to define. Because of the radiation problem, waveguides, as a rule, are not terminated in open circuits. However, open-circuit terminations are used frequently for coaxial lines, with radiation being prevented by extending the outer conductor beyond the abruptly cut inner. This extended shield acts as a waveguide-below-cutoff attenuator for the frequencies at which the coaxial line is used in the TEM mode. The effect of fringing fields near the end of the inner conductor can be represented by a discontinuity capacitance in the end plane. These fields cause the effective position of the open circuit to be displaced away from the end of the inner conductor by an amount computed by Somlo [54], and shown in Fig. 3.12 for both 7 mm and 14 mm outer diameter lines.

3.3.1.3 Intermediate value reflection coefficient standards: Intermediate value reflection coefficient standards with $0 < |\Gamma| < 1$ are useful for calibrating reflectometers and checking their linearity. Any device having an intermediate value Γ of reflection coefficient (variable or fixed) of sufficient stability may be used as a transfer standard, provided that Γ was measured by some independent means. There are many commercially available coaxial terminations consisting of resistors, which terminate a line with a specified VSWR. These devices are not absolute standards in the sense that the reflection coefficient they produce is not calculable from basic physical units, but nevertheless are extremely useful as transfer standards.

Absolute intermediate reflection standards may be constructed with little more effort than matched standards, by connecting two lines of different characteristic impedance, using the principle outlined in Section 3.2.2. For example, the authors have constructed an intermediate reflection standard for 14 mm outer diameter

* In the authors' experience, one particular commercially available sliding 'short circuit' is inadequately damped behind the shorting element, and behaves as a matched load at some frequencies and positions within its specified operating range!

coaxial line by using a 14 mm diameter outer conductor, an inner conductor of approximately 10·8 mm diameter, and a beadless connector. The resulting characteristic impedance of approximately 17 ohms gives $|\Gamma| \simeq 0{\cdot}5$ when used to terminate a 50 ohm system. The standard is equipped with a sliding load (tapered lossy dielectric cone), which is used in the same manner as discussed in Sections 3.2.1 and 3.3.1.1 for a sliding matched load. The reflection coefficient of the standard was calculated after taking into account the step capacitance at the connector plane [8].

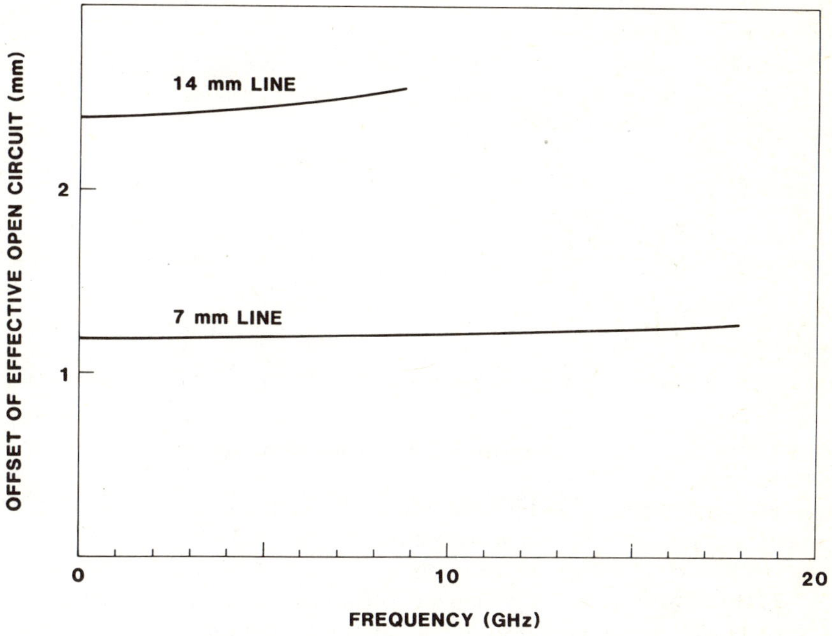

Fig. 3.12. *Displacement of the effective position of coaxial shielded open circuits for 7 mm and 14 mm outer diameter 50 ohm coaxial lines, as a function of frequency*

Reduced height waveguide intermediate reflection standards are available. They also incorporate a sliding load, and the value of Γ can be calculated using eqn. 3.6, and taking into account the step capacitance [55].

For both coaxial and waveguide stepped characteristic impedance standards, the effect of step capacitance is small compared to, and is in quadrature with, the value of Γ calculated from the characteristic impedance discontinuity.

The most recent design suggestion for a variable intermediate reflection coefficient standard in waveguide is the use of an offset waveguide junction [23]. When a precision waveguide incorporating a sliding load is offset at its flange from a guide of the same dimensions, the reflection coefficient is calculable, as discussed in Section 3.1.2.

3.3.2 Two-ports

Any linear reciprocal two-port may be characterized by the three S parameters S_{11}, S_{22}, and $S_{12} = S_{21}$ representing the input and output reflection coefficients and the transfer coefficient, respectively, associated with the same port impedance Z_0. The transfer coefficients for some reflectionless two-ports ($S_{11} = S_{22} = 0$) are $\exp(-\gamma l)$ for a length of uniform transmission line, $\exp(-j\beta)$ for an ideal phase shifter with no attenuation, and $\exp(-\alpha)$ for an ideal attenuator with no phase shift. An ideal (lossless) impedance transformer designed to provide a reflectionless connection between impedances Z_1 and Z_2 has transfer parameters $S_{12} = 1/S_{21} = (Z_1/Z_2)^{1/2}$.

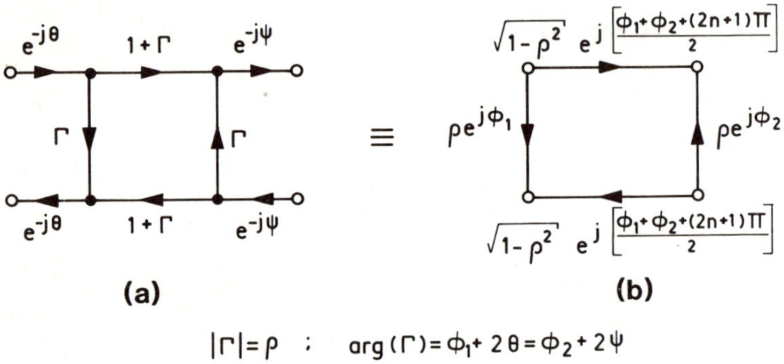

Fig. 3.13. *Two equivalent flow graph representations of a lossless reciprocal two-port*

Two equivalent flow graph representations of a lossless reciprocal two-port with the same designated port impedances are shown in Fig. 3.13.

3.3.2.1 Tuners: Adjustable impedance transformers, or tuners, are often used in microwave circuits to either transform an impedance into a match (no reflections), or into a complex conjugate impedance for maximum power transfer.

In precision microwave measurement systems, it is sometimes necessary to provide impedance matches at particular ports which are difficult to access by a reflectometer. In such cases, a tuner–isolator combination forms a convenient unit which may be removed from the circuit and adjusted to present a match at the tuner port. When this portable unit is returned to the circuit, any mismatched termination of the isolator port has negligible effect on the preadjusted tuner port match, provided the isolator has sufficient isolation (say > 40 dB).

An important property of a tuner is the range of impedances it is capable of matching to the transmission line system for which it was designed. Consider a tuner which has been adjusted so that an impedance Z_0^* is presented at port 1 when impedance Z_L^* is connected to port 2.† It follows from the theory of transmission

† A superscript * denotes the complex conjugate.

line junctions that if the tuner is lossless, the impedance seen looking into port 2 is Z_L when port 1 is terminated in Z_0. Consequently, if a lossless tuner is capable of matching any passive impedance Z_L to Z_0, then it must also be capable of presenting any passive impedance at port 2 when Z_0 is connected to port 1. For Z_0 real, a restriction on Z_L is that it must have a non-zero real part, because it is not possible for a lossless tuner terminated in a purely reactive load to dissipate the energy associated with a real impedance. Therefore a lossless tuner is incapable of providing a match to a Z_0 which has a real part, if $Z_L = jX_L$, which lies on the perimeter of the Smith chart.

3.3.2.2 Sliding-stub tuner: The input admittance of a length of lossless transmission line of characteristic admittance Y_0 which is terminated in an admittance Y_L can be matched to Y_0 by connecting in parallel with the line a short-circuited length of line (a short-circuited stub). The circuit is depicted in Fig. 3.14*a*. The tuning or matching is effected by adjusting the position of the stub along the line, and the position of the short circuit in the stub.

The normalized admittance seen looking towards the termination from the junction J is

$$Y' = Y_s' + Y_{in}' \tag{3.29}$$

where, from eqn. 1.15,

$$Y_s' = -j \cot (\beta l_s); \quad Y_{in}' = \frac{Y_L' + j \tan (\beta l)}{1 + j Y_L' \tan (\beta l)} \tag{3.30}$$

and it is clear that Y_s' is always reactive. If the stub is slid along the line until l is such that $Y_{in}' = 1 + jB'$, and l_s is varied until $Y_s' = -jB'$, then $Y' = 1$ and the line is matched.

This matching task is illustrated on the Smith chart of Fig. 3.14*b*, normalized to Y_0. As we move along the transmission line away from Y_L' the admittances seen looking towards the termination trace, in a clockwise direction, a circular arc concentric with the centre of the Smith chart. When the angle subtended at 0 is θ, this arc intersects the unit circle at A, where $Y_{in}' = 1 + jB'$. If the stub tuner is attached at A, adjustment of the position of the short circuit can cause the sum Y' of the parallel admittances to trace the arc of the unit circle from A to 0, where $Y' = 1$. Since one revolution around the Smith chart corresponds to an electrical movement of $\lambda_g/2$, where λ_g is the wavelength of the propagation in the transmission line, the distance from the load to A is $\theta\lambda_g/4\pi$. The length l_s of the adjusted stub can also be determined from the Smith chart by proceeding in a clockwise direction along the perimeter through an angle ϕ from the point C, representing a short circuit, to the point representing $-jB'$. Then $l_s = \phi\lambda_g/4\pi$.

The practical difficulties associated with sliding one section of parallel-connected transmission line along another are such that, except for two-wire transmission lines, a telescopic section, termed a 'line stretcher', must be used to alter the distance between the load and a stub tuner, thereby achieving the effect of sliding

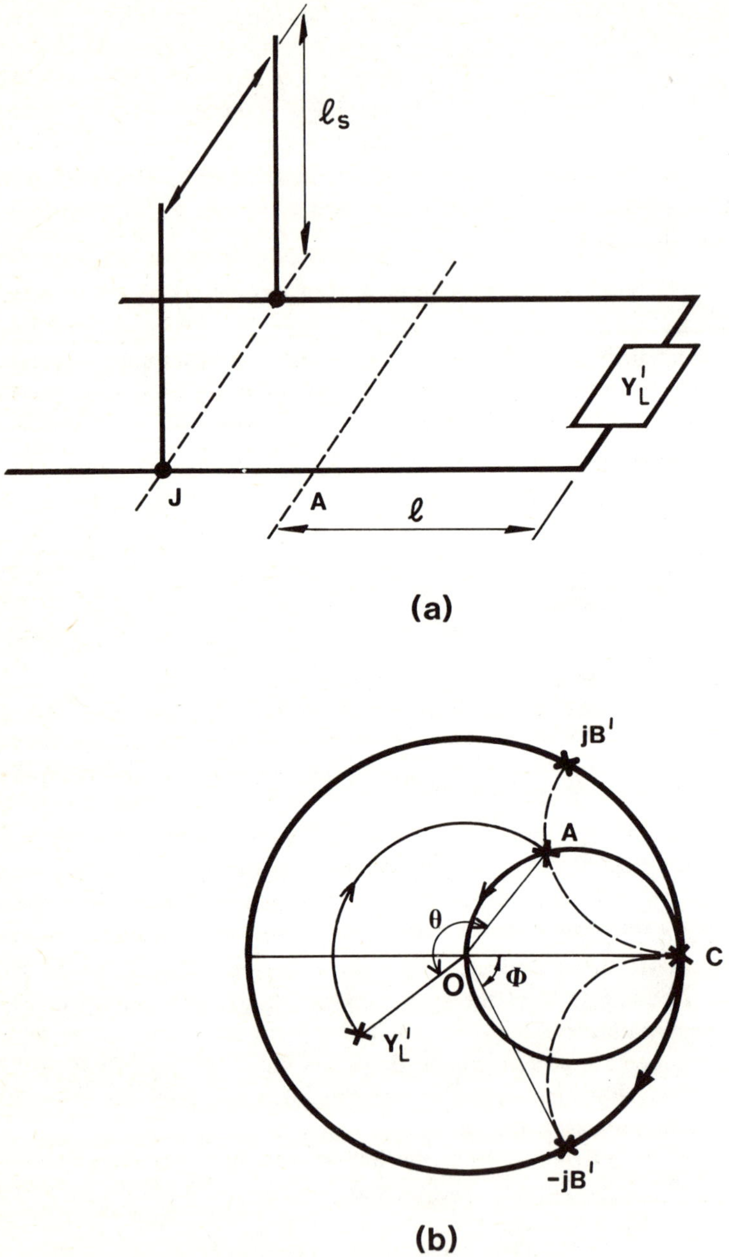

Fig. 3.14. *Sliding-stub tuner*
a Schematic representation in a circuit
b The matching procedure illustrated on the Smith chart

the stub. However, problems can occur with sliding contacts usually associated with the line stretcher.*

3.3.2.3 Slide-screw tuner and stub stretcher: A waveguide slide-screw tuner, as shown in Fig. 3.15*a*, is similar in concept to the sliding-stub tuner. A metallic post is inserted into the waveguide to provide susceptance tuning, and the location of the post is made adjustable along a longitudinal slot. Although the variation of the susceptance of the post with penetration does not cover the full susceptance range, as does a stub tuner, all capacitive susceptances are obtainable. Therefore the post is positioned longitudinally so that a capacitive post susceptance will provide the necessary match. The length of slot which is required in order that any impedance may be matched by the tuner is therefore longer than if the full range of susceptances were available from adjustment of the post penetration.

Another commercially available tuner, the coaxial stub stretcher, is shown in Fig. 3.15*d*, and schematically in Fig. 3.16*a*, and is similar in concept to the sliding-stub tuner. A stub tuner which is paralleled to the main transmission line has an inner conductor which also acts as the outer conductor of a concentric coaxial line. Both the stub and the concentric line are terminated in sliding short circuits.

Unfortunately, this compact arrangement is incapable of presenting a match at port 1 for all possible passive loads at port 2. When port 1 is terminated in a (real Z_0) match, the matched admittance at plane A is shunted by the susceptance of the stub terminated by the short circuit at P. The composite admittance lies on the unit circle of the admittance Smith chart in Fig. 3.16*b* at a point such as Y. If Fig. 3.16*b* is now transformed into an impedance Smith chart, the unit circle transforms into the dashed circle, and an admittance such as Y images into an impedance Z. The coaxial line terminated by the short circuit at S is in series with the main transmission line, and adjustment of the position of S will vary the series reactance, causing the impedance seen from port 2 to move along a circle of $\mathrm{Re}(Z) = \text{constant}$. Therefore, with a matched load at port 1, no adjustment of the stub stretcher will present at port 2 an impedance which lies inside the unit circle of the (impedance) Smith chart. From the result of Section 3.3.2.1, it follows that the stub stretcher cannot present a match at port 1 when the load impedance lies inside the unit circle. In such circumstances, a transmission line with a length near $\lambda_g/4$ should be interposed between load impedance and the stub stretcher, so that the impedance at port 2 of the line stretcher is near to being diametrically opposite the load impedance on the Smith chart, and therefore outside the unit circle.

3.3.2.4 Double-stub tuner: A coaxial double-stub tuner is shown schematically in Fig. 3.17*a* as two coaxial lines terminated in sliding short circuits and connected

* Because of their construction, simple line stretchers do not have a constant characteristic impedance, but more complicated line stretchers with constant characteristic impedance have been made.

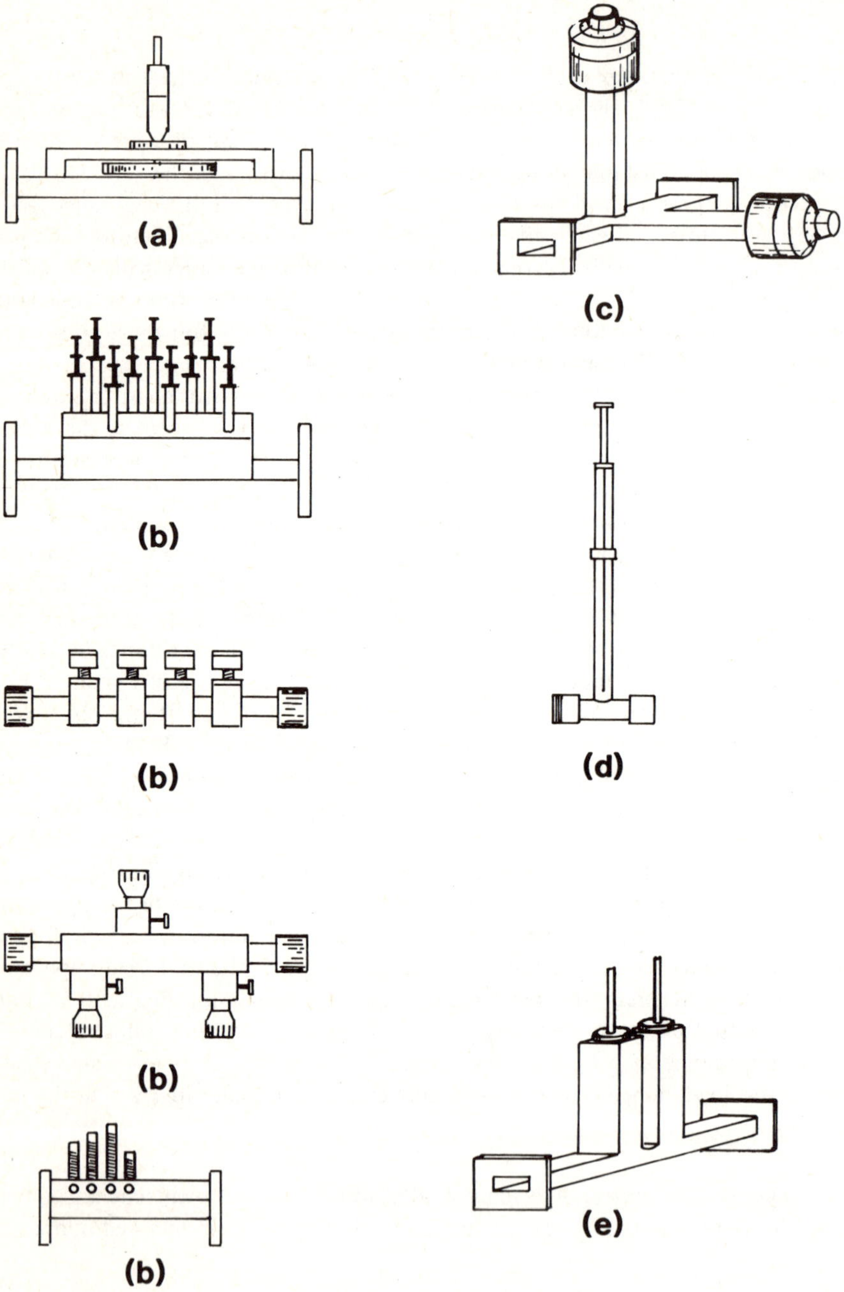

Fig. 3.15. *Various realizations of coaxial and waveguide tuners*

in parallel with the main transmission line a distance s apart. The operation of the double-stub tuner is illustrated on the admittance Smith chart of Fig. 3.17*b*. We know that provided the admittance at plane A of the transmission line lies on the unit circle, the susceptance can be adjusted using stub 1 until the input admittance at port 1 is a match. Admittance values on the unit circle transform through an angle $\theta = 4\pi s/\lambda_g$ to become admittance values on the circle shown dashed in Fig. 3.17*b*. It is the function of stub 2 to modify the susceptance presented at plane B until the admittance at plane B lies on this transformed circle.

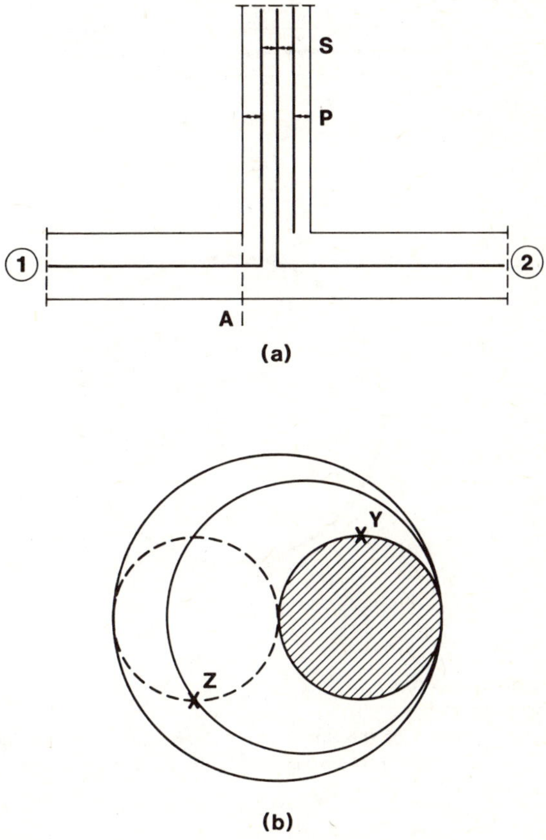

Fig. 3.16. *Coaxial stub stretcher*
a Schematic diagram
b The hatched region of the Smith chart shows the domain of impedances which cannot be transformed into a match using a coaxial stub stretcher

Consequently, any admittance at plane B which lies inside the hatched circle cannot be tuned to a match by the double-stub tuner.

This limitation of the tuning range of a double-stub tuner is, in principle, less restrictive the smaller s becomes. However, in practice, values of s near multiples of $\lambda_g/2$ increase the sensitivity of the stub adjustment, and a more favoured approach is to build the tuner such that the distance between port 1 and stub 1 differs from that between port 2 and stub 2 by $\lambda_g/4$, with $s = \lambda_g/4$. In this case, any admittance which is not initially tuneable is readily tuned after the double-stub tuner has been reversed in the circuit.

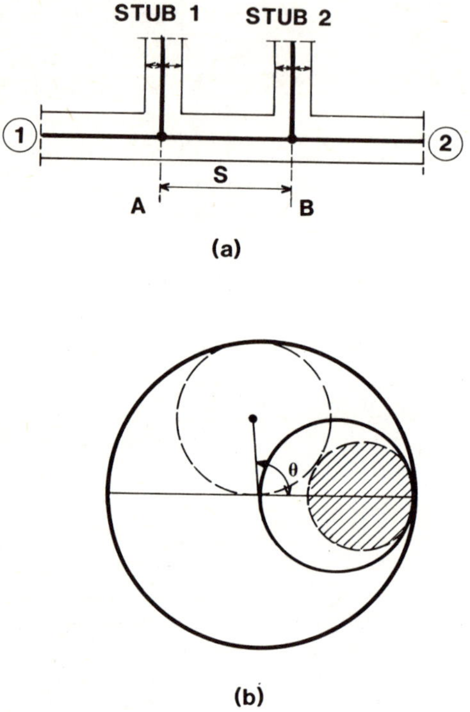

Fig. 3.17. *Coaxial double-stub tuner*
a Schematic diagram
b The hatched region of the Smith chart shows the domain of admittances which cannot be transformed into a match using the double-stub tuner. θ is related to the stub separation s by $\theta = 4\pi s/\lambda_g$

3.3.2.5 E-H tuner: The *E-H* tuner shown in Fig. 3.15*c* and schematically in Fig. 3.18*a* is the result of creating *E*-plane and *H*-plane junctions at a common transverse plane of the main transmission guide. The junction is intrinsically a four-port identical to the magic tee discussed in Section 3.3.4.1, but with adjustable short circuits in the transverse arms it may be regarded as a two-port. By a suitable choice of the reference planes, the S parameters of the *E-H* tuner are given by

$$S_{11} = S_{22} = \tfrac{1}{2}(e^{-j\phi} + e^{-j\psi}), \quad S_{12} = S_{21} = \tfrac{1}{2}(e^{-j\phi} - e^{-j\psi})$$

where ϕ, ψ are the phases of the reflection coefficients from the short circuits in the E-plane and H-plane arms, respectively.

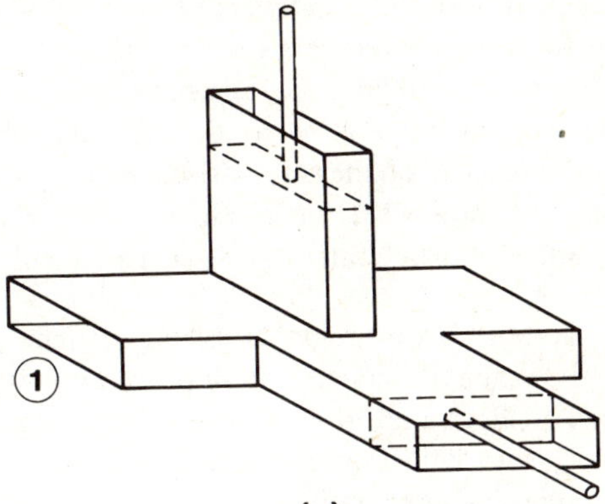

(a)

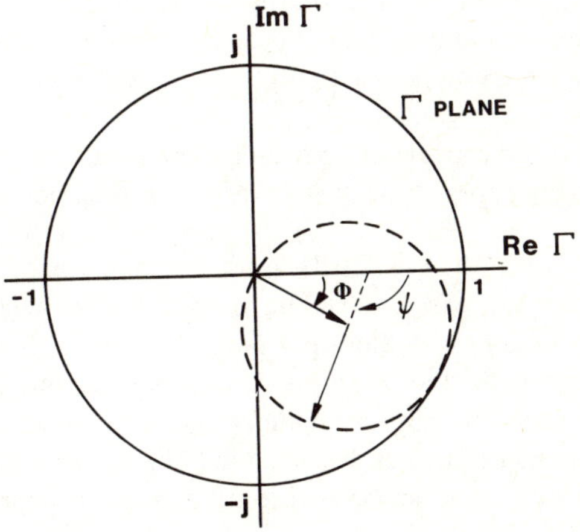

(b)

Fig. 3.18. *Waveguide E-H tuner*
a Schematic diagram
b The Γ plane diagram showing that the *E-H* tuner can match any passive impedance which is not purely reactive

In order to determine the tuning range of the device, we note that the reflection coefficient $\Gamma_1 = S_{11}$ at port 1 when port 2 is terminated in a match can be located on the complex Γ plane as shown in Fig. 3.18*b*. Each of the two component vectors has a magnitude of $\frac{1}{2}$, and the resultant can lie anywhere within the circle

$\Gamma = 1$, which defines the perimeter of the Smith chart. Invoking the result of Section 3.3.2.1, it follows that the *E-H* tuner can be used to match to any passive impedance (with, of course, the exception of purely reactive loads). In the past, the range of this matching ability has been underestimated [56].

Examination of Fig. 3.18*b* reveals two interesting properties of an *E-H* tuner. First, when the impedance at port 1 is a match, adjustment of the two short-circuit positions will provide orthogonal movements on the Γ plane only if the magnitude of the reflection coefficient of the load is $1/\sqrt{2}$. Secondly, if the magnitude of the load reflection coefficient is near zero or unity, the two component vectors of Γ_1 on the complex Γ plane are near colinearity. Consequently, the closer the load impedance is to a match, the more difficult it is to adjust the *E-H* tuner to present a match at port 1, because the short-circuit adjustments are highly interdependent.

3.3.2.6 Multi-stub tuners: Multi-stub tuners, consisting of lengths of transmission line with several dielectric or metallic posts whose penetrations can be individually varied, take many forms, as illustrated in Figs. 3.15*b*. In waveguides, the posts are usually designed to provide variable capacitive susceptance. In coaxial lines, the slight withdrawal of a large diameter metallic post from the outer conductor can give rise to additional inductive susceptance [5].

In order to reduce the size of multi-stub tuners, but to retain the ease of admittance adjustment, lumped inductance may be introduced between the stubs in the form of transmission line with an impedance higher than the characteristic impedance.

Multi-stub tuners are capable of providing a broadbanded impedance match. The tuning ranges of many types of multi-stub tuner are described in the literature [57].

3.3.2.7 Tuning for a match: One of the most frequently encountered requirements of a tuner is that it be adjusted so that the termination connected to its output port is transformed into a match at the input port of the tuner. Many of the impedance measuring systems described in this book can be used to determine when the tuning has been completed successfully, by comparing the impedance presented with that of an impedance standard such as discussed in Section 3.2.1. However, the ease and speed with which the tuner can be successfully adjusted depends on the measurement system used.

The slotted lines discussed in the next chapter have, for many years, been the most commonly available impedance measuring apparatus. However, their use to follow the progress of a tuning adjustment is laborious, and the tuning becomes more or less a trial and error procedure which is continued until the VSWR falls below an acceptable minimum value.

Far more rapid tuner adjustment is possible if the measurement system provides a display of both the real and imaginary parts of the reflection coefficient, as on a Smith chart. Such a display is obtainable from many of the reflectometer measurement systems to be discussed in the following chapters, or from fixed multi-probe methods as discussed in Section 4.2.3.1. The high resolution reflectometer discussed in Section 5.4.3 allows rapid tuner adjustment by minimizing the multiple reflections between the tuner port and a fixed reflecting object.

A special waveguide tuning aid was described by Hollway and Somlo [58], and called the match checker. It can be inserted between waveguide ports which are each required to be matched, and the tuners associated with each port can be adjusted *in situ*. The match checker consists of a plain section of waveguide in which a bilaterally (nominally) matched attenuator, in the form of a tapered resistive load, may be moved axially. A reflective object also in the waveguide can be moved to either end of the attenuator, and locked to it. When the reflecting object is moved to one end of the attenuator and locked to it, the transmission through the system will vary as the attenuator is slid inside the waveguide. This variation is the result of multiple reflections between the match checker and the mismatched port facing the reflecting object. The tuner associated with this port is adjusted to minimize these variations. The reflecting object is then moved to the other end of the attenuator, and the process repeated for the second tuner. After a few repetitions of this process, both ports will be matched. The match checker is suitable for use only with measuring systems which are capable of resolving small changes in the transmitted signal level.

3.3.3 Three ports

The most often encountered three-ports are tee junctions such as the coaxial tee, and the E-plane and H-plane tee junctions shown with an H-plane Y junction in Figs. 3.19a, b, c, and d, respectively. The sliding-stub tuner considered in Section 3.3.2.2 is intrinsically a three-port although it may be regarded as a two-port when the third port is inaccessible for external connection.

Three important theorems associated with lossless reciprocal three-ports supporting one mode of propagation at one frequency are [53]

1. A position can be found for a short circuit in one arm such that there is no energy transmission between the other two arms.
2. If a three-port is symmetrical about an arm containing a short circuit, a position of the short circuit can be found for which there is complete transmission between the other two arms without loss or reflection.
3. A lossless reciprocal three-port cannot present three matched ports simultaneously.

It follows from Theorem 3 that a three-port which presents three matched ports simultaneously is either lossy and/or non-reciprocal, such as a circulator.

3.3.3.1 E-plane and H-plane junctions: The H-plane junction is similar to the coaxial tee in that, provided the terminations are not close to the waveguide junction, the arms are electrically in parallel, while for the E-plane junction the arms are electrically in series.

If a tuner is included in the symmetrical arm of an H-plane or E-plane junction and adjusted so that $S_{33} = 0$, and the reference planes in arms 1 and 2 are

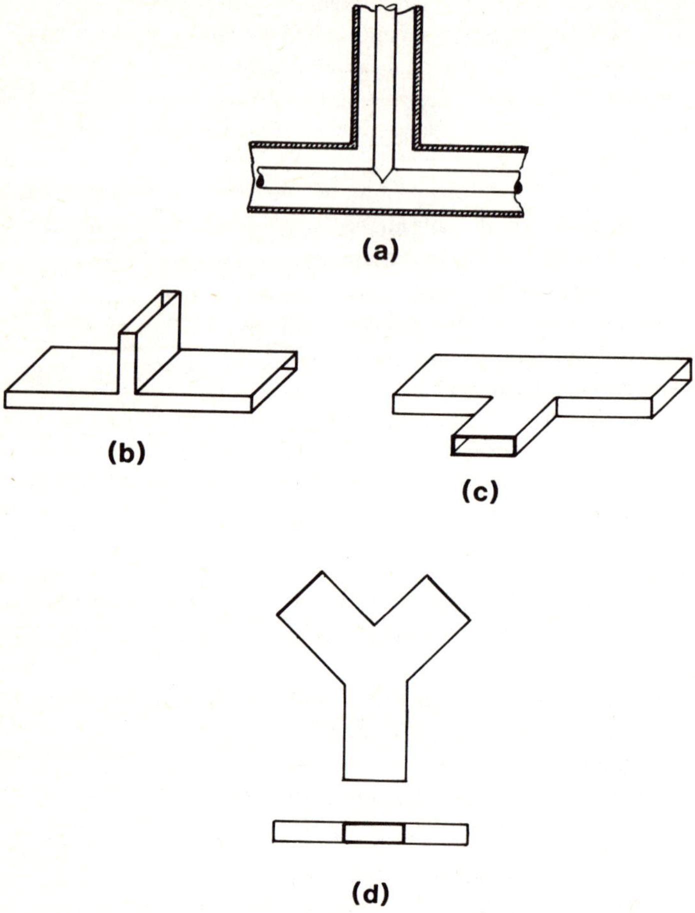

Fig. 3.19. *Various forms of three-port junctions*

chosen so that all the elements of the $\boldsymbol{S}$ matrices are real, then for the H-plane junction [53]

$$\boldsymbol{S} = \begin{bmatrix} -\frac{1}{2} & \frac{1}{2} & \frac{1}{\sqrt{2}} \\ \frac{1}{2} & -\frac{1}{2} & \frac{1}{\sqrt{2}} \\ \frac{1}{\sqrt{2}} & \frac{1}{\sqrt{2}} & 0 \end{bmatrix} \tag{3.31}$$

and for the E-plane junction

$$S = \begin{bmatrix} \frac{1}{2} & \frac{1}{2} & \frac{1}{\sqrt{2}} \\ \frac{1}{2} & \frac{1}{2} & -\frac{1}{\sqrt{2}} \\ \frac{1}{\sqrt{2}} & -\frac{1}{\sqrt{2}} & 0 \end{bmatrix} \quad (3.32)$$

Notice that in accordance with Theorem 3 a simultaneous match at all three ports has not been achieved. However, either junction may be used as an efficient power splitter since $|S_{13}| = |S_{23}| = \frac{1}{\sqrt{2}}$.

3.3.3.2 Circulators (isolators): An ideal three-port circulator is a non-reciprocal lossless junction, simultaneously matched on all ports and with scattering parameters given by

$$S = \begin{bmatrix} 0 & 0 & 1 \\ 1 & 0 & 0 \\ 0 & 1 & 0 \end{bmatrix} \quad (3.33)$$

for the direction of circulation indicated in Fig. 3.20*a*. The non-reciprocal properties of passive circulators are obtained by using non-reciprocal elements such as magnetically biased ferrite materials. By reversing the bias polarity, the direction of circulation may be reversed, and the device may then be used as an electronically operated microwave switch (Section 3.4).

If one arm of a three-port circulator is internally terminated in a matched load, the two-port device obtained is termed an isolator, and is usually denoted by one of the symbols in Fig. 3.20*b*. The scattering parameters of an ideal isolator are

$$S = \begin{bmatrix} 0 & 0 \\ 1 & 0 \end{bmatrix} \quad (3.34)$$

For an imperfect isolator $S_{12} \neq 0$, and a measure of the quality of the isolator is the ratio of the forward to reverse transmission. This ratio, termed the isolation I, is given in decibels by

$$I = 20 \log_{10} [|S_{21}|/|S_{12}|] \quad (3.35)$$

3.3.4 Four-ports

The most common four-ports encountered in microwave measurement systems are the hybrid junctions, which ideally are lossless, reflectionless, reciprocal four-ports having the property that energy injected into one port will divide between two of the remaining three ports. Therefore, all hybrid junctions have directional properties. The energy division is assumed to be equal unless stated otherwise.

Hybrid junctions may be classified by the phase difference of the signals appearing at the output ports. A quadrature (90°) hybrid junction is commonly called a directional coupler. In the following sections both 180° and 90° hybrid junctions will be discussed.

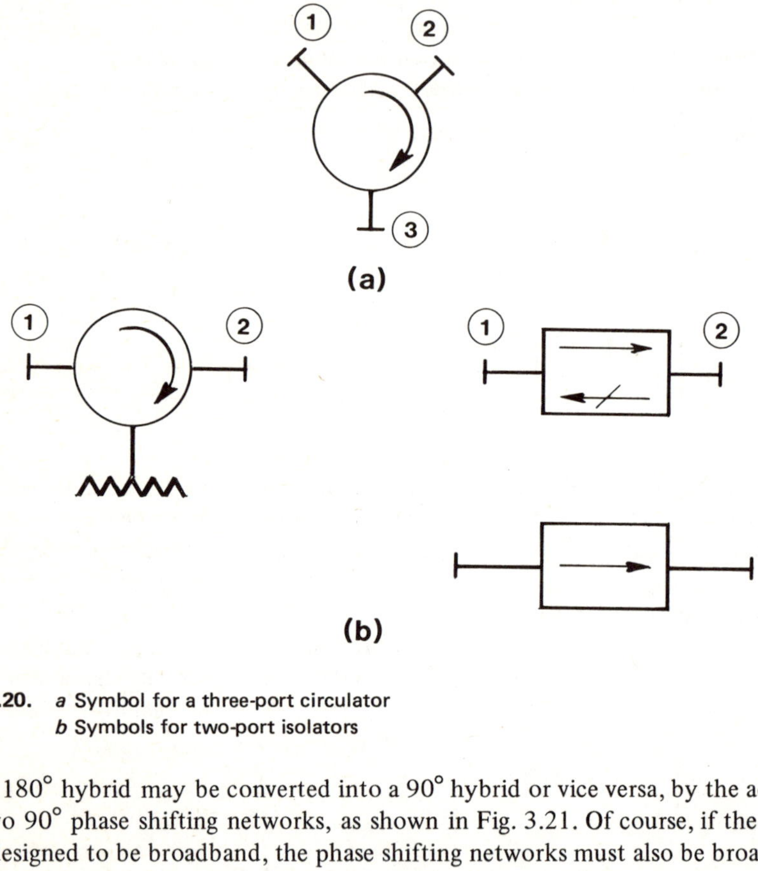

Fig. 3.20. *a* Symbol for a three-port circulator
b Symbols for two-port isolators

A 180° hybrid may be converted into a 90° hybrid or vice versa, by the addition of two 90° phase shifting networks, as shown in Fig. 3.21. Of course, if the hybrid was designed to be broadband, the phase shifting networks must also be broadband.

3.3.4.1 180° hybrid, magic tee: A microwave 180° hybrid exhibits behaviour similar to that of the low-frequency hybrid transformer, shown in Fig. 3.22*a*, which is most remembered for its use in telephony to isolate the microphone from the earpiece. 180° hybrids are realized as discrete transformers up to about 500 MHz, and at higher frequencies as either TEM stripline structures [59], or in waveguides, symmetrical junctions or asymmetrical ring circuits [60]. In rectangular waveguides, the symmetrical junction form of the 180° hybrid is known as the 'magic tee'.

A 180° hybrid may be adapted for many uses including applications as a tuner, power divider, summer, differencer, phase shifter, and with non-linear elements, a multiplier. When the hybrid appears in one of these guises, it is often referred to by the name of the function for which it has been adapted. For example, we have already seen that an appropriately terminated magic tee is called an *E-H* tuner, and power dividers are often constructed using a 180° hybrid.

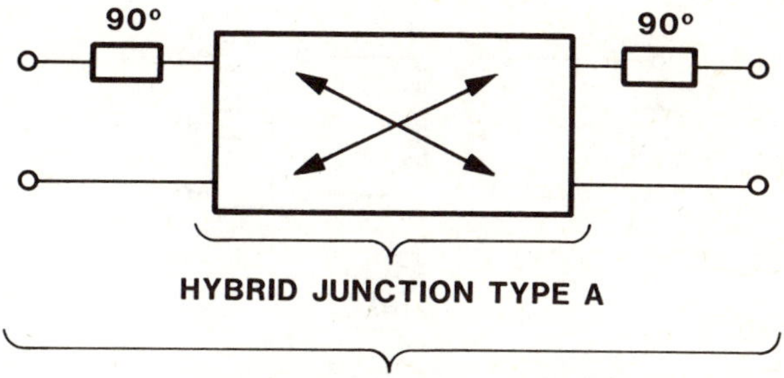

Fig. 3.21. *Conversion of a 90° hybrid junction to a 180° hybrid junction, or vice versa, using two 90° phase-shifting networks*

The S parameters of the magic tee in Fig. 3.22*b* are

$$\mathbf{S} = \frac{1}{\sqrt{2}}\begin{bmatrix} 0 & 0 & 1 & 1 \\ 0 & 0 & -1 & 1 \\ 1 & -1 & 0 & 0 \\ 1 & 1 & 0 & 0 \end{bmatrix} \tag{3.36}$$

It follows that if all ports are matched and a signal a_4 is applied to port 4, it appears equally divided at ports 1 and 2. In this guise the 180° hybrid is a power splitter.

Similarly, if a signal a_3 is applied at port 3, it appears equally divided but in antiphase at ports 1 and 2. Therefore if a_3 and a_4 are applied simultaneously, the outputs b_1 and b_2 at ports 1 and 2 are

$$b_1 = \frac{1}{\sqrt{2}}(a_4 + a_3)$$

$$b_2 = \frac{1}{\sqrt{2}}(a_4 - a_3)$$

In this guise, the 180° hybrid has an important role as an equality detector (if $a_3 = a_4$, $b_2 = 0$). If the outputs of square-law detectors attached to ports 1 and 2 are differenced, a quantity proportional to $a_3 a_4$ is obtained, and in this guise the 180° hybrid is a multiplier.

When the magic tee is in the *E-H* tuner configuration discussed in Section 3.3.2.5, and shown schematically in Fig. 3.18*a*, it is apparent from Fig. 3.18*b* that there is an infinite number of positions related by $\psi = \phi + \pi$ into which the pair of tuning stubs can be set to cause a match to be presented at port 1 when port 2 is terminated in a match. Then $S_{12} = S_{21} = e^{-j\phi}$, and the magic tee may be used as a phase shifter.

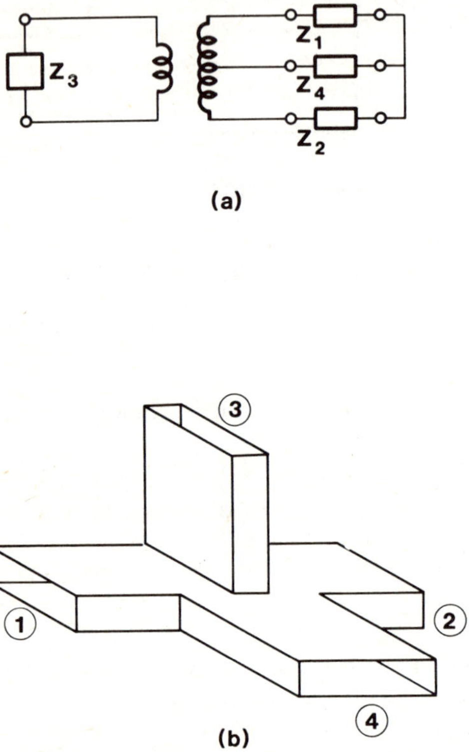

Fig. 3.22. *180° hybrid junctions*
a Low-frequency realization
b Waveguide realization – magic tee

The practical challenge in the manufacture of a 180° hybrid is to make the junction usable over a wide frequency range. Decade bandwidth has been reported in stripline construction [59], and in waveguides the bandwidth of a magic tee can be increased by introducing irises or posts into the junction region, or by using ridged waveguides. When the distributed transformer in Fig. 3.23 was introduced into the magic tee junction shown, the resulting junction exhibited side arm isolation of more than 20 dB over the 9–12 GHz band, and more than 15 dB from

8 GHz to 12 GHz.* The corresponding isolation of the uncompensated junction varies from 10 dB to 3 dB over this frequency range.

3.3.4.2 90° hybrid, directional coupler: The coupling C of the lossless reciprocal 90° hybrid (directional coupler) in Fig. 3.24*a* is the ratio (in decibels) of the incident power at port 2 to the power delivered to a matched termination at port 3, with other ports terminated in reflectionless impedances. When the real, designated port impedances are identical, C is related to the coupling coefficient c by

$$C = -20\log_{10}(c) = -10\log_{10}(|S_{32}|^2) \tag{3.37}$$

Fig. 3.23. ***Photograph of a waveguide magic tee with broadbanding component (distributed transformer) removed from the junction***

Similarly, the directivity D is the ratio (in decibels) of the power coupled out on port 3 for a given incident power on port 2, to that out of port 3 with the same incident power on port 1. The directivity is therefore a measure of the ability of the coupler to selectively couple power flowing in one direction in the main arm. The directivity is related to the directivity coefficient d by

$$D = 20\log_{10}(d) = 10\log_{10}(|S_{32}/S_{31}|^2) \tag{3.38}$$

The scattering parameters of the ideal directional coupler are

* P. I. Somlo

$$
S = \begin{bmatrix} 0 & \sqrt{(1-c^2)} & 0 & -jc \\ \sqrt{(1-c^2)} & 0 & -jc & 0 \\ 0 & -jc & 0 & \sqrt{(1-c^2)} \\ -jc & 0 & \sqrt{(1-c^2)} & 0 \end{bmatrix} \tag{3.39}
$$

If a signal a_1 is applied at port 1 of a directional coupler with all ports matched, it

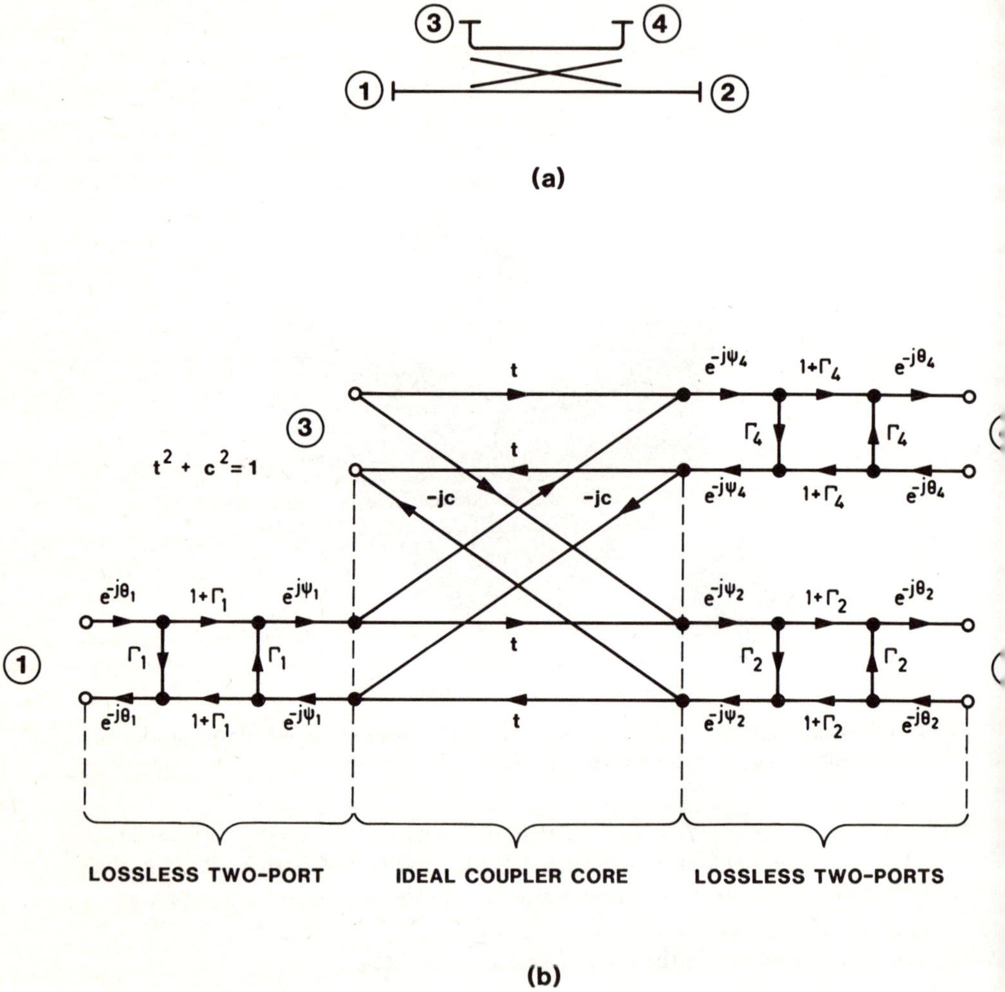

Fig. 3.24. *A 90° reciprocal, lossless, hybrid junction i.e. a directional coupler*
a Symbol
b Flow graph representation

appears at ports 2 and 4 divided such that b_4 is C dB less than a_1, and with a 90° phase difference between b_2 and b_4.

A 90° (quadrature) hybrid with $c = 1/\sqrt{2}$ is a 3 dB directional coupler. Consequently, a signal a_1 appears equally divided at ports 2 and 4, with a 90° phase difference.

A directional coupler may be synthesized by the tandem connection of two directional couplers. In particular, a 3 dB coupler results from the tandem connection of two couplers having $c = [(2-\sqrt{2})/4]^{1/2}$ ($C = 8{\cdot}34321$ dB) [61]. It has been analytically and experimentally verified that the tandem connection of two 3 dB couplers using a fixed and a variable phase shifter can synthesize a coupler of (theoretically) any coupling [62, 63].

Practical directional couplers may take many forms. Up to frequencies of several hundred megahertz lumped-element realizations are possible, and distributed coupling may be used down to 1 MHz, where the necessary electrical and magnetic coupling is obtained by coiling a twisted (insulated) pair of wires into a helix [64]. At microwave frequencies couplers are realized either as coupled TEM structures, as single intrinsically directive coupling apertures between waveguides [65], or as a set of coupling apertures $\lambda_g/4$ apart at midband between the common broadwall of parallel waveguides [66].

A non-ideal lossless reciprocal coupler may be modelled as an ideal coupler with (lossless) tuners attached to any three of its four ports [67]. The flow graph of such a coupler is shown in Fig. 3.24*b*. Either of the flow graphs in Fig. 3.13 may be used to represent the tuners. In so far as a practical coupler is lossless and reciprocal, this flow graph and the use of the flow graph reduction techniques enable the properties of the practical coupler to be studied with various terminations on the four ports [68]. In particular, it follows from the above result that a practical (non-ideal) lossless coupler may be converted into an ideal coupler by attaching correctly adjusted tuners to three of its four ports.

If port 4 of the lossless, reciprocal coupler in Fig. 3.24*a* is internally terminated in a matched load, the S parameters of the resulting three-port are

$$\boldsymbol{S} = \begin{bmatrix} 0 & \sqrt{(1-c^2)} & 0 \\ \sqrt{(1-c^2)} & 0 & -jc \\ 0 & -jc & 0 \end{bmatrix} \tag{3.40}$$

3.3.4.3 Use of directional couplers as levelling components: A directional coupler is widely used as the sampling element in a closed-loop feedback system designed to maintain at a constant level the signal incident on the load. A typical circuit is shown in Fig. 3.25*a*.

The flow graph of a practical coupler whose fourth port is terminated in a fixed reflection is shown in Fig. 3.25*b*, and using flow graph reduction techniques it can be shown that the transmission from port 1 to port 3 is

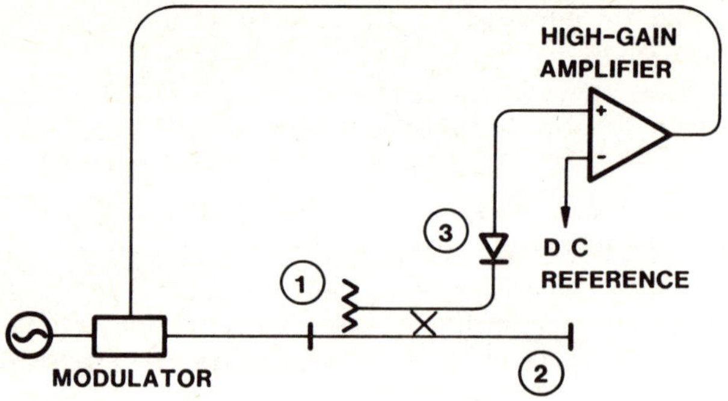

(a)

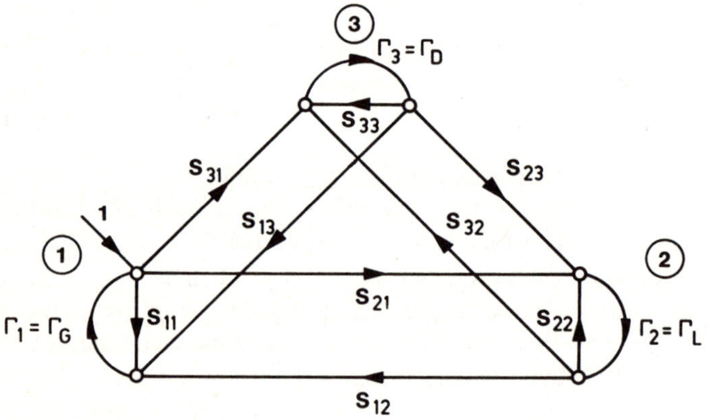

(b)

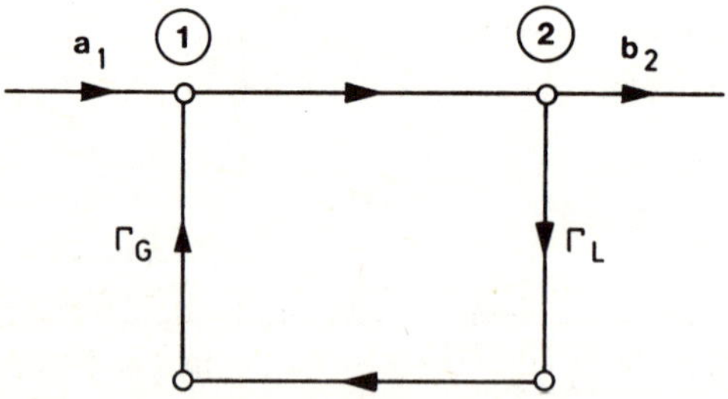

(c)

Fig. 3.25. *a* Simple circuit using a directional coupler as a levelling component
b Flow graph representation of a directional coupler with the fourth port internally terminated, a generator connected to port 1, a load to port 2, and a detector to port 3
c Flow graph representation of a generator connected directly to a load

$$\frac{b_3}{a_1} = [S_{31} + \Gamma_2(S_{21}S_{32} - S_{31}S_{22})]/[1 - S_{11}\Gamma_1 - S_{22}\Gamma_2 - S_{33}\Gamma_3 + \Gamma_1\Gamma_2(S_{11}S_{22} - S_{21}S_{12}) + \Gamma_1\Gamma_3(S_{11}S_{33} - S_{31}S_{13}) + \Gamma_2\Gamma_3(S_{22}S_{33} - S_{32}S_{23}) + \Gamma_1\Gamma_2\Gamma_3(S_{11}S_{32}S_{23} + S_{22}S_{31}S_{13} + S_{33}S_{21}S_{12} - S_{11}S_{22}S_{33} - S_{21}S_{32}S_{13} - S_{31}S_{23}S_{12})] \tag{3.41}$$

The symmetry of Fig. 3.25*b* allows the interchange of subscripts 2 and 3 in eqn. 3.41, and hence after forming the ratio b_2/b_3

$$b_2 = b_3\left[\frac{S_{21}}{S_{31}} + \Gamma_D\left(S_{23} - \frac{S_{21}S_{33}}{S_{31}}\right)\right] + a_2\left(S_{22} - \frac{S_{21}S_{32}}{S_{31}}\right) \tag{3.42}$$

If b_3 is held constant by the levelling circuitry, the signal emerging from port 2 is constant, except for the effect of the second term in eqn. 3.42. The magnitude of this term depends on the quality of the coupler, since S_{22} is the coupler's output port reflection coefficient, and $|S_{31}/S_{32}|$ is a measure of its directivity. In practice, only $|b_3|$ is held constant, and therefore the phase of the signal incident on the load, when measured with respect to a decoupled phase reference inside the generator, varies with ϕ_L. Because impedance measurements are made with respect to the phase at the measurement plane, this variation is inconsequential.

For best results, the level detector should be temperature stable, and have low $1/f$ type noise characteristics. When PIN diodes are used as modulating elements, their S-shaped characteristics provide higher loop gain near the midpoint of the modulation range than can be achieved near the maximum and minimum levels.

When the circuit in Fig. 3.25*a* is used with a frequency swept source, the output level will vary because of the frequency response of the coupler. To calculate this variation, we assume the coupler is ideal, and define the normalized power P_1' to be the maximum value of c^2 within the swept frequency range. Thus P_1' is the maximum value of normalized power delivered to a matched detector connected to port 3. The level variation of the output signal at port 2 is then R, where

$$R = \frac{1}{\alpha}\frac{(1 - \alpha P_1')}{(1 - P_1')} \tag{3.43}$$

and α is the coupling reduction factor. (If the coupling decreases by 1 dB from P_1' at some point in the frequency range, then at that frequency $\alpha = 10^{-0\cdot 1} = 0{\cdot}794$.) It follows from eqn. 3.43 and the curves in Fig. 3.26 that R(dB) is always greater than the coupling reduction.

An improvement in output levelling is obtained if a second coupler with an identical frequency response is introduced to form the three-port network, which has been termed 'super levelling' [69], as shown in Fig. 3.27. The level variation of the output signal at port two is

$$R = \frac{1}{\alpha}\frac{(1 - \alpha P_1')(1 - P_2')}{(1 - P_1')(1 - \alpha P_2')} \tag{3.44}$$

where P_2' is the maximum value of c^2 for the second coupler. An examination of the curves in Fig. 3.27 shows the best levelling is obtained when the second coupler has large coupling, and the first has small coupling, which also increases the power available at the measuring port.

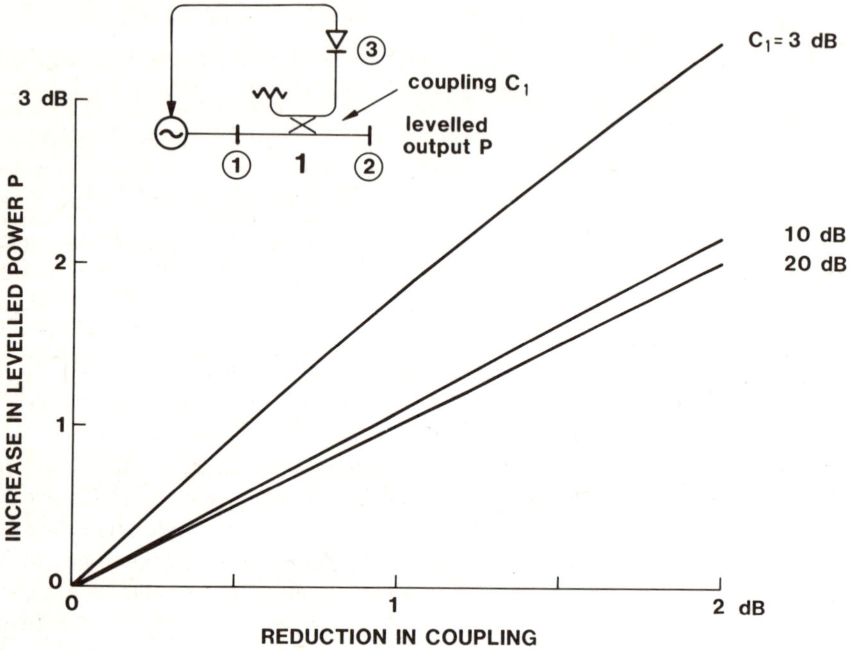

Fig. 3.26. *Wideband performance of a simple levelling circuit*
The variation of the levelled output power *P* is shown as a function of the variation of the coupling coefficient

When sufficient power is available from the generator, the three-port constructed from two couplers with identical frequency responses, and shown in Fig. 3.28, gives

$$R = \frac{1-\alpha P_1'}{1-P_1'} \tag{3.45}$$

The output variation is independent of the coupling of the second coupler, although this would normally be chosen to be a nominal 3 dB coupler to ensure that as much power as possible was available at the measuring port.

An 'ultra-flat' levelling network is obtained if a third coupler with identical frequency response is introduced into the circuit of Fig. 3.28 to give the three-port shown in Fig. 3.29. The level variation of the output signal at port 2 is

$$R = \frac{(1-\alpha P_1')(1-P_2')}{(1-P_1')(1-\alpha P_2')} \tag{3.46}$$

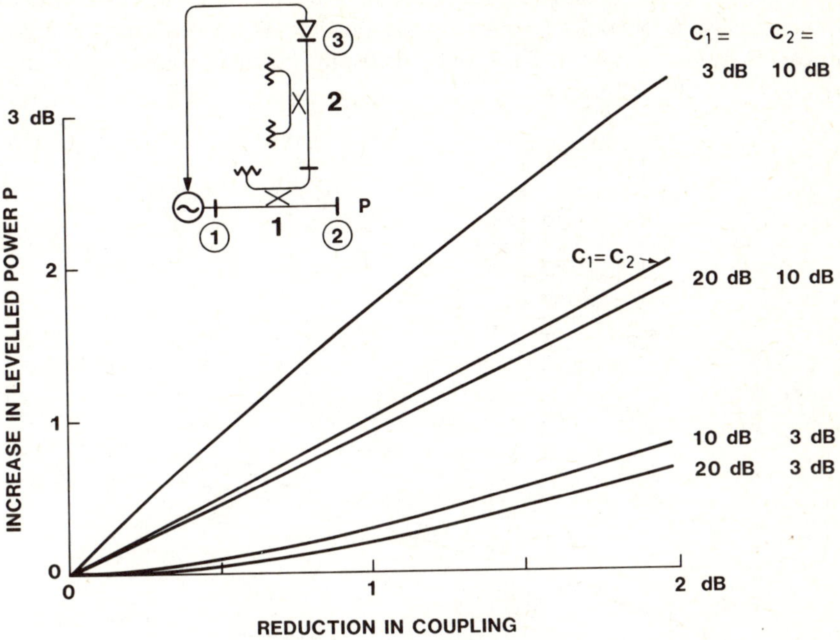

Fig. 3.27. *Wideband performance of a super levelling circuit*
The variation of the levelled output power *P* is shown as a function of the variation of the coupling coefficient

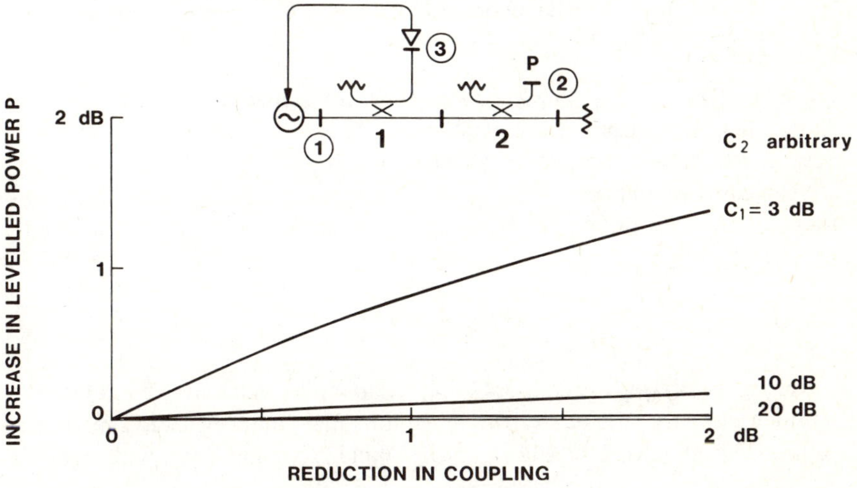

Fig. 3.28. *Wideband performance of a tandem levelling circuit*
The variation of the levelled output power *P* is shown as a function of the variation of the coupling coefficient

which is independent of the coupling of the third coupler, although as before this would normally be chosen to be as large as possible. When couplers 1 and 2 are identical, $R = 0$ dB, but even if they have different but small coupling the output variations are small, as shown in the figure.

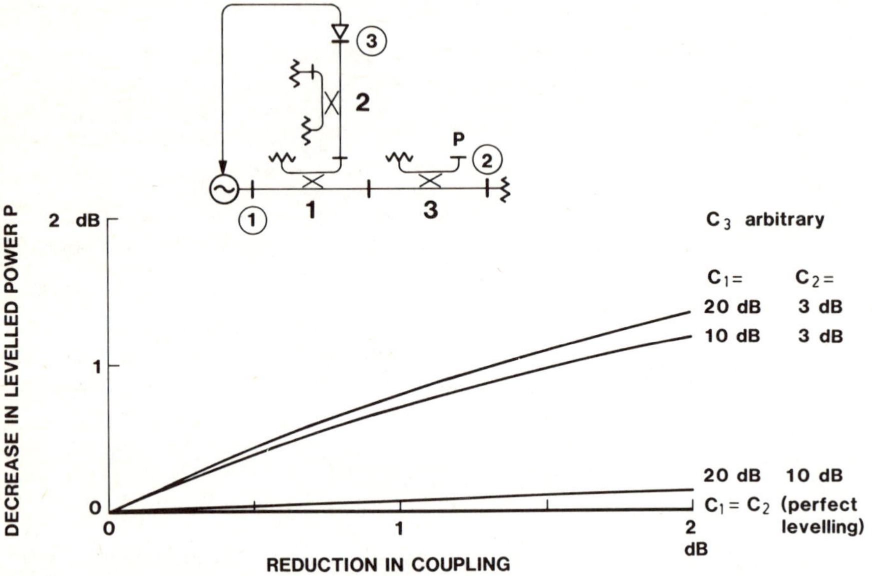

Fig. 3.29. *Wideband performance of an ultra-flat levelling circuit*
The variation of the output power *P* is shown as a function of the variation of the coupling coefficient

3.3.4.4 Use of directional couplers to improve the generator match: Apart from providing a levelled output, the circuit shown in Fig. 3.25*a* can also improve the generator match as seen from the measuring port [70].

If a generator is connected directly to a termination, the signal b_2 incident on the termination when the generator output is a_1 can be found from the flow graph of Fig. 3.25*c* to be

$$b_2 = \frac{a_1}{1 - \Gamma_G \Gamma_L} \tag{3.47}$$

After writing $a_2 = \Gamma_L b_2$ in eqn. 3.42 and assuming b_3 is maintained constant by the levelling circuitry, the equivalent generator reflection coefficient presented to the termination at port 2 of Fig. 3.25*a* is found by comparison with eqn. 3.47 to be

$$\Gamma_e = S_{22} - \frac{S_{21} S_{32}}{S_{31}} \tag{3.48}$$

which is solely a function of coupler quality, and independent of the generator impedance.

If the coupler is of good quality, Γ_e may be a considerable improvement over Γ_G. For example, if a 10 dB coupler ($|S_{21}| = 0{\cdot}95$) having a directivity of 40 dB ($|S_{31}/S_{32}| = 100$) and with $|S_{22}| = 0{\cdot}01$ is used in the levelling loop, the maximum value of $|\Gamma_e|$ is 0·0195.

Of course, if the generator is itself well-matched, the use of an imperfect coupler in a levelling loop may result in $|\Gamma_e|$ being larger than $|\Gamma_G|$.

3.4 Switches

Modern microwave switches are either of the mechanical or solid-state type, and are classified according to the number of their ports and possible interconnections, and to the number of stable states (latching or not). The three main microwave parameters governing switch performance are insertion loss, isolation, and residual reflection (VSWR).

Solid-state switches reduce the microwave signal by either reflecting or absorbing the incident signal. They are controlled by a d.c. bias, and operate with the solid-state elements in a fully saturated condition. With few exceptions, these switches are built in TEM lines with coaxial connectors, and are available commercially with insertion losses of less than 2 dB, and isolations of greater than 80 dB. The absorptive (PIN) diodes, which are used saturated in solid-state switches, may also be used unsaturated as level modulating elements in feedback circuits such as in Fig. 3.25*a*. In the past, solid-state switches used ferrite material, either in the form of a circulator with a reversible bias to change the direction of circulation, or as a Faraday-rotation switched isolator. Nowadays the use of ferrite switches is confined mainly to high-power applications.

Mechanical switches are available for both coaxial line and waveguide systems. Coaxial mechanical switches use make and break signal carrying contacts and have a limited lifetime ($\sim 10^6$ operations). Their switching speeds (typically $\sim$ 30 ms) are slower than solid-state switches, but they offer low insertion loss ($<$ 0·5 dB at 18 GHz is realisable). Waveguide mechanical switches are mostly of the 'gas-tap' type, with the signal entering through a port in an outer 'stator', and then passing through a waveguide tunnel in an inner 'rotor', before exiting through a second stator port. Manual or electrical rotor rotation effects the switching by changing the tunneling configuration between the various stator ports. The better quality waveguide switches of this type do not rely on mechanical contact for signal transfer between the stator and the rotor, and some exhibit exceptional performance with insertion loss of less than 0·05 dB, isolation greater than 90 dB, and VSWR less than 1·01. More importantly, for precision measurements, the scatter of the insertion loss on repeated switchings can be less than 0·001 dB. Somlo and Hollway [71] utilized this high repeatability by switching in or out of circuit an additional loss path to produce a transfer-standard attenuator, as shown in Fig. 3.30. A highly repeatable phase increment can be produced using the same principle, by switching in additional signal path length.

One of the most important measurement applications of microwave switching is the 'Dicke-switch' [72], used in the comparison of the levels of two weak signals buried in receiver noise. The receiver input is switched repeatedly between the unknown signal and a known reference signal (time-division multiplexing), and the detected receiver output is synchronously detected in a phase-sensitive detector (PSD, lock-in amplifer). Dicke switches find extensive application in radio astronomy and the calibration of microwave noise sources.

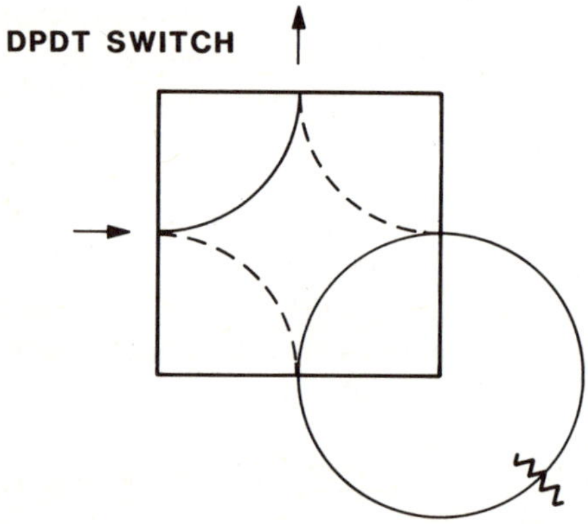

Fig. 3.30. *A highly repeatable attenuation step built using a mechanical double-pole double-throw waveguide switch*

References

1 MACKENZIE, T. E. and SANDERSON, A. E.: 'Some fundamental design principles for the development of precision coaxial standards and components', *IEEE Trans.*, 1966, **MTT-14**, pp. 29–39

2 ENGEN, G. F.: 'An extension to the sliding short method of connector and adaptor evaluation', *J. Res. NBS*, 1971, **75C**, pp. 177–183

3 SKILTON, P. J.: 'Loss, reflection and repeatability in waveguide flanged couplings', *CPEM Digest*, 1974, pp. 271–273

4 SKILTON, P. J.: 'A technique for measuring the efficiency of waveguide-to-coaxial-line adaptor', *IEEE Trans.*, 1978, **IM-27**, pp. 231–234

5 DOBBIE, L. G. and SOMLO, P. I.: 'Precise absolute technique for measuring low-loss symmetrical linear passive 4-terminal networks at V.H.F. and U.H.F.', *Proc. IEE*, 1963, **110**, pp. 1171–1180

6 'IEEE standard for precision coaxial connectors', *IEEE Trans.*, 1968, **IM-17**

7 WHINNERY, J., JAMIESON, H. and ROBBINS, T.: 'Coaxial line discontinuities', *Proc. IRE.*, 1944, **32**, pp. 695–709

8 SOMLO, P. I.: 'The computation of coaxial-line step capacitances', *IEEE Trans.*, 1967, **MTT-15**, pp. 48–53

9 JURKUS, A.: 'Computation of step capacitances in coaxial line', *IEEE Trans.*, 1972, **MTT-20,** pp. 708–709

10 GREEN, H.E.: 'Numerical solution of some important transmission-line problems', *IEEE Trans.*, 1965, **MTT-13,** pp. 676–692

11 THURSTON, W. R.: 'A radically new connector for the laboratory', *Gen. Rad. Experim.*, 1948, **23**

12 WOODS, D.: 'A coaxial connector system for precision r.f. measuring instruments and standards', *Proc. IEE*, 1961, **108,** pp. 205–213

13 SANDERSON, A. E.: 'A radically new coaxial connector for high-precision measurements', *Gen. Rad. Experim.*, 1963, **37**

14 'Calibration standards for precision coaxial lines', *Gen. Rad. Experim.*, 1966, **40**

15 ADAM, S. F., KIRKPATRICK, G. R., SLADEK, N. J. and BRUNO, S. T.: 'A new 34 GHz, 3·5 mm low-cost utility coaxial connector featuring low leakage, low standing-wave ratio, and long life', *IEEE Trans.*, 1976, **MTT-24,** pp. 995–997

16 BRUNO, S. T. and KIRKPATRICK, G. R.: 'A significant improvement over SMAs in test applications', *Microwave J.*, 1979, **22,** pp. 67–69

17 ERWIN, J. B.: 'Mating and performance of N-type connectors', Telecom Australia Res. Labs. Rep R6, 1979, 7281

18 GRIFFIN, E. J., and HARRIS, I. A.: 'Calculation of capacitance of adjacent steps in parallel-plane transmission lines', *Proc. IEE*, 1976, **123,** pp. 729–733

19 WHINNERY, J. R. and JAMIESON, H. W.: 'Equivalent circuits for discontinuities in transmission lines', *Proc. IRE*, 1944, **32,** p. 98

20 FRANZ, K.: 'Reflection of Hl waves at a sudden change of the cross section of a rectangular wave-guide', *Frequenz*, 1948, **2,** pp. 227–231

21 MILES, J. W.: 'The equivalent circuit for a plane discontinuity in a cylindrical waveguide', *Proc. IRE*, 1946, **34,** p. 728

22 WHEELER, H. A. and SCHWIEBERT, H.: 'Step-twist waveguide components', *IRE Trans.*, 1955, **MTT-3,** p. 45

23 HUNTER, J. D.: 'The displaced rectangular waveguide junction and its use as an adjustable reference reflection', *IEEE Trans.*, 1984, **MTT-32,** pp. 387–394

24 JOHNSON, R. H. and HARVEY, D.: 'Hydraulic waveguide flange clamp for precision measurements', *Electron. Lett.*, 1983, **19,** pp. 730–731

25 HOLLWAY, D. L. and SOMLO, P. I.: 'The reduction of errors in a precise microwave attenuator calibration system', *IEEE Trans.*, 1973, **IM-22,** pp. 268–270

26 PAKAY, P. and TOROK, A.: 'Koaxialis csatlakozok beiktatasi csillapitasa reprodukalodasanak vizsgalata', *Meres es Automatika*, 1977, **25,** pp. 219–222

27 NEMOTO, T.: 'The measurement of a waveguide joint properties', *Electr. & Comm. Japan*, 1969, **52-B,** pp. 68–76

28 SOMLO, P. I.: 'The effect of flange loss on the reflection coefficient of reduced-height waveguide reflection standards', *IEEE Trans.*, 1979, **MTT-27,** pp. 795–797

29 BEATTY, R. W. and KERNS, D. M.: 'Recently developed microwave impedance standards and methods of measurement', *IRE Trans.*, 1958, **I-7,** pp. 319–321

30 WARNER, F. L.: 'Microwave attenuation measurement' (Peter Peregrinus, 1977)

31 WEINSCHEL, B. O.: 'Air-filled coaxial lines as absolute impedance standards', *Microwave J.*, 1964, 7, pp. 47–50

32 HARRIS, I. A. and SPINNEY, R. E.: 'The realization of high frequency impedance standards using air-spaced coaxial lines', *IEEE Trans.*, 1964, **IM-13,** pp. 265–272

33 WOODS, D.: 'Immittance transformation using precision air-dielectric coaxial lines', *Proc. IEE*, 1971, **118,** pp. 1667–1674

34 LACY, P. and OLDFIELD, W.: 'Calculable physical impedance references in automated precision reflection measurement', *CPEM Digest*, 1980, pp. 330–335

35 WOODS, D.: 'Coaxial 2-port resistor and its application to r.f. measurements and standards', *Proc. IEE*, 1971, **118,** pp. 1859–1866

36 FARMER, E. D.: 'Junction admittance between waveguides of arbitrary cross-sections' (IEE Monograph, 148R, 1955)

37 BEATTY, R. W.: '2-port quarter-wave waveguide standard of voltage standing wave ratio', *Electron. Lett.*, 1973, **9**, pp. 24–26

38 KERNS, D. M. and GRANDY, Jun., W. T.: 'Perturbation theorems for waveguide junctions with applications', *IEEE Trans.*, 1966, **MTT-14**, pp. 85–92

39 MAGID, M.: 'Precision microwave phase shift measurements', *IRE Trans.*, 1958, **IM-7**, pp. 321–331

40 KERNS, D. M. and HEDBERG, R. W.: 'Propagation constant in rectangular waveguide of finite conductivity', *J. Appl. Phys.*, 1954, **25**, pp. 1550–1551

41 SCHAFFELD, W. and BAYER, H.: 'Uber des Verhalten elektromagnetischer Wellen in Kreiszylindrischen Hohlleitern im Bereich der Grenzenfrequenz unter Berucksichtung des endlichen Wandleitfahligkeit', *Arch. Elektr. Ubertragung*, 1956, **10**, pp. 89–97

42 RAMO, S., WHINNERY, J. R. and VAN DUZER, T.: 'Fields and waves in communication electronics' (Wiley and Sons, 1965), pp. 51–53

43 CLARRICOATS, P. J. B. and SLINN, K. R.: 'Numerical solution of waveguide discontinuity problems', *Proc. IEE*, 1967, **114**, pp. 878–886

44 MASTERMAN, P. H. and CLARRICOATS, P. J. B.: 'Computer field-matching solution of waveguide transverse discontinuities', *Proc. IEE*, 1971, **118**, pp. 51–63

45 HOWE, G. W. O.: 'The application of telephone transmission formulae to skin-effect problems', *J. IEE*, 1916, **54**, pp. 473–480

46 SOMLO, P. I.: 'The computation of surface impedance of multilayer cylindrical conductors' CSIRO (Australia) Nat. Meas. Lab. Rep, 1966, no. APR 12

47 SOMLO, P. I.: 'Recession depth in metallic conductors at low frequencies', *Electron. Lett.*, 1971, **7**, pp. 776–777

48 WHEELER, H. A.: 'Formulas for skin effect', *Proc. IRE*, 1942, **30**, pp. 412–424

49 ZORZY, J.: 'Skin effect correction in immittance and scattering coefficient standards employing precision air-dielectric coaxial lines', *IEEE Trans.*, 1966, **IM-15**, pp. 265–272

50 NELSON, R. E. and CORYELL, M. R.: 'Electrical parameters of precision, coaxial, air-dielectric transmission lines', (NBS Monograph, 1966, no. 96)

51 FARADAY PROCTOR, R.: 'High frequency resistance of plated conductors', *Wireless Eng.*, 1943, **20**, p. 56

52 KERNS, D. M. and HEDBERG, R. W.: Private communication

53 KERNS, D. M. and BEATTY, R. W.: 'Basic theory of waveguide junctions and introductory microwave network analysis' (Pergamon Press, 1967).

54 SOMLO, P. I.: 'The discontinuity capacitance and the effective position of a shielded open circuit in a coaxial line', *Proc. IREE* (*Australia*), 1967, **28**, pp. 7–9

55 MARCUVITZ, N. (ed.): 'Waveguide handbook' (McGraw-Hill 1951), Chap. 5

56 HARVEY, A. F.: 'Microwave engineering' (Academic Press, 1963), pp. 120–121

57 RAGAN, G. C.: 'Microwave transmission circuits' (McGraw-Hill, 1948)

58 HOLLWAY, D. L. and SOMLO, P. I.: 'The match checker – a simple instrument for matching impedances in waveguide and coaxial systems', *IEEE Trans.*, 1974, **MTT-22**, pp. 560–561

59 ARAIN, M. H. and SPENCER, N. W.: 'Tapered asymmetric microstrip magic tee', *IEEE Trans.*, 1975, **MTT-23**, pp. 1064–1067

60 MONTGOMERY, C. G., DICKE, R. H. and PURCELL, E. M.: 'Principles of microwave circuits' (McGraw-Hill, 1948) Chap. 9.13

61 HINDIN, H. J. and ROSENZWEIG, A.: '3-dB couplers constructed from two tandem connected 8·34 dB asymmetric couplers', *IEEE Trans.*, 1968, **MTT-16**, pp. 125–126

62 TEETER, W. L. and BUSHORE, K. R.: 'A variable ratio microwave power divider and multiplexer', *IRE Trans.*, 1957, **MTT-5**, pp. 227–229

63 VAILLANCOURT, R. M.: 'Analysis of the variable ratio microwave power divider', *IRE Trans.*, 1958, **MTT-6**, pp. 238–239

64 SOMLO, P. I.: 'Precision impedance measurement and Smith-chart display with easily fabricated directional couplers from 1 MHz upwards', *Proc. IREE* (*Australia*), 1974, **35**, pp. 341–345

65 BETHE, H. A.: 'Theory of diffraction by small holes', *Phys. Review*, 1944, **66**, p. 163

66 MILLER, S. E. and MUMFORD, W. W.: 'Multi-element directional couplers', *Proc. IRE*, 1952, **40**, p. 1071

67 KAHN, W. K. and KYHL, R. L.: 'The directional coupler core of an arbitrary, lossless, reciprocal 4-port', *Proc. IRE*, 1961, **49**, pp. 1699–1700

68 SOMLO, P. I. and MORGAN, I. G.: 'A side-arm switched directional coupler as a single-step attenuator of high long-term stability', *IEEE Trans.*, 1974, **MTT-22**, pp. 830–835

69 'Microwave power measurement' Hewlett-Packard Application Note 64, 1965

70 ENGEN, G. F.: 'Amplitude stabilization of a microwave signal source', *IRE Trans.*, 1958, **MTT-6**, pp. 202–206

71 SOMLO, P. I. and HOLLWAY, D. L.: 'The Australian National Standards Laboratory X-band radiometer for the calibration of noise sources', *IEEE Trans.*, 1968, **MTT-16**, pp. 664–669

72 DICKE, R. H.: 'The measurement of thermal radiation at microwave frequencies', *Rev. Sci. Instr.*, 1946, **17**, pp. 268–275

Chapter 4

Detection of the sum of incident and reflected waves

The nature of the voltage variation along a uniform transmission line supporting a single mode of propagation at a single frequency and terminated in a passive load was discussed in Section 1.3.2. The use of the results of that section in determining the reflection coefficient of the termination is the subject of this chapter.

4.1 Basic principles

The major results of Section 1.3.2 will be restated here with one important alteration. Instead of the distance from the termination being denoted by z, which is negative between the termination and the generator, it will be denoted by l, which is defined by

$$l = -z \tag{4.1}$$

and is therefore zero at the termination, and increasingly positive as the line is traversed towards the generator. The latter definition is found to be more convenient in practice than the former.

The magnitude of the voltage on the line is, from eqn. 1.23

$$|V| = [1 + |\Gamma_L|^2 + 2|\Gamma_L| \cos(2\beta l - \phi_L)]^{1/2} \tag{4.2}$$

and is a maximum wherever $2\beta l - \phi_L = 2n\pi$, and a minimum wherever $2\beta l - \phi_L = (2n-1)\pi$, $n = 0, 1, 2, \ldots$. The voltage standing wave ratio S, and the magnitude of the terminating reflection coefficient Γ_L remain related by

$$\frac{|V_{max}|}{|V_{min}|} = S = \frac{1+|\Gamma_L|}{1-|\Gamma_L|}, \quad |\Gamma_L| = \frac{S-1}{S+1} \tag{4.3}$$

It can be seen from Fig. 1.3 that $|V|$ varies more rapidly in the vicinity of minima than maxima. The reason for this is apparent if the locus of $|V|$ with l is considered in the geometrical representation of eqn. 4.2 on the Smith chart in Fig. 4.1. The closer $|\Gamma_L|$ is to unity, the more rapidly $|V|$ varies in the vicinity of a minimum. Fig. 4.1 also illustrates graphically that if $|\Gamma_L| = 1$, the minimum value of $|V|$ is zero.

Because of the more rapid variation near minima than maxima, it is usually the minima of a standing wave pattern which are sought in order to determine ϕ_L. However, if the standing wave detector responds to $|V|^2$ rather than $|V|$, then it will register

$$|V|^2 \propto 1 + |\Gamma_L|^2 + 2|\Gamma_L| \cos(2\beta l - \phi_L) \tag{4.4}$$

and this sinusoidal behaviour describes equally rapid variations in the vicinity of the maxima and minima. This will be the case if a power meter, or in particular a semiconductor diode at low power levels (square law region), is used as the detector.

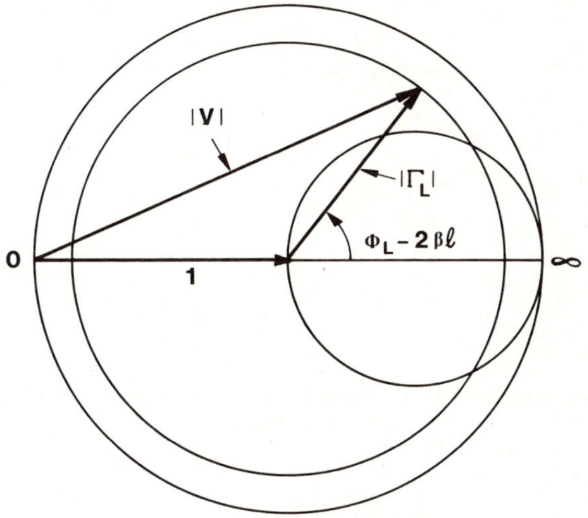

Fig. 4.1. *Variation of |V| along a uniform, lossless, mismatched transmission line, as shown on an impedance Smith chart*

In practice, it is rare to determine the distance l_L of a minimum in the voltage standing wave from the termination Γ_L. It is more convenient to determine the separation between l_L, and the location l_S of a voltage minimum when the termination is a short circuit ($\Gamma = e^{j\pi}$) as illustrated in Fig. 4.2. The value of ϕ_L is then obained from

$$\phi_L + 2n\pi = \pi\left[1 - \frac{4}{\lambda_g}(l_S - l_L)\right] \tag{4.5}$$

where λ_g is the wavelength in the transmission system, and the integer n is chosen to ensure that ϕ_L falls within the desired principal range.

There are several advantages in using a short-circuit termination as a phase reference in this manner. Short-circuit terminations are comparatively easy to manufacture in either coaxial or waveguide systems, and a mechanical inspection is usually all that is necessary to define the electrical reference plane $l=0$. Furthermore, in so far as the short circuit and the line are lossless, the voltage standing minima are zero.

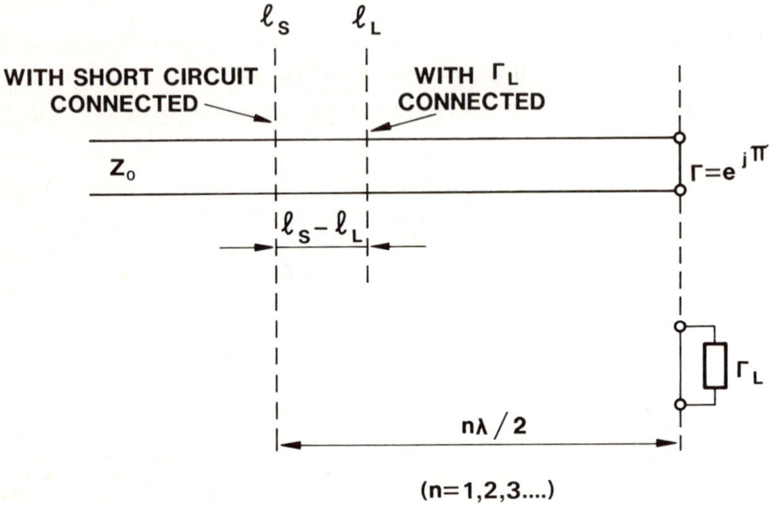

Fig. 4.2. *Locations of voltage minima along a transmission line when a reference short circuit or a mismatched termination* Γ_L *is connected*

4.2 Methods using magnitude detectors

Measurement methods which determine the impedance of a termination by the use of detectors responding only to $|V|$ or $|V|^2$ may be classified as either single-probe or multiprobe.

The first category involves the use of a single probe to scan the voltage standing wave pattern. This may be achieved either by moving the probe along the transmission line, or by causing the standing wave pattern to pass by a fixed probe. In either case, mechanical movement is necessary.

The second category involves inferring the terminating impedance from the responses of several probes fixed at intervals along the line. Although measurement methods in this category avoid the need for mechanical movement, they are constrained to operate over a limited frequency range because of the fixed separation between probes.

The first reported systematic investigation of standing waves on a transmission line was by Professor Lecher [1] in 1890, who described the detection (with fluorescent tubes) of standing waves along a pair of stretched, parallel wires. Lecher's objective was not to measure impedance, but rather the velocity of propagation along the line. In 1931 Labus [2] measured the impedance of a termination by using a pair of parallel wires $3\lambda/4$ long with current detectors (thermal convertors) at each end.

In order to monitor with high repeatability the voltage (or current) distribution along a transmission line without causing significant electrical loading, non-contacting methods of sampling the electric (or magnetic) field are usually employed. In practice, one of two types of detection is used. Either the microwave signal is brought out for heterodyne detection, or a detector is incorporated in the probe assembly, which may or may not be tuned. Typical circuits using these detection methods are shown in Fig. 4.3.

If the transmission line does not have an open structure such as the twin-wire line, strip line, surface or dielectric waveguide, or the slab line shown in Fig. 4.4*a*, the outer conductor must be slotted to gain access to the field inside.

4.2.1. Slotted lines

Slotted lines are usually used to measure reflection coefficients of small or moderate magnitude. Reflection coefficients having a magnitude near unity are difficult to measure using a slotted line, not only because the voltage minima are near zero, and can disappear into the noise of the detector, but also because the dynamic range of the measurement is large.

Slotted lines are used most commonly for coaxial systems operated in the TEM mode, and for rectangular waveguide systems operated in the fundamental TE_{10} mode, as shown in Fig. 4.4*b, c*. For minimum disturbance to the field, the slot is best cut along a surface line where the surface currents flow only in the axial direction. It is shown in Fig. 4.5 that the wall currents for a rectangular waveguide operating in the TE_{10} mode are axial only on the centre of the broad wall, which is therefore the preferred location of a slot.

Barlow and Cullen [3] described in detail the mechanical construction of slotted lines. For precision measurements, the main considerations of the design are:

(*a*) the slotted line should have a well defined characteristic impedance, close to the nominal value;
(*b*) the slot-effect should be small, it should not change the characteristic impedance significantly, and should not cause significant reflections;
(*c*) the probe coupling should have minimal variations along the line and should have high long-term stability;
(*d*) the probe penetration should be capable of being varied, and have a facility for locking;
(*e*) the line should be equipped with good quality connectors, particularly on the measuring port where high repeatability is even more important than low reflection;
(*f*) the slotted line should have low losses.

Early investigators such as Oliver [4] and Sorrows *et al.* [5] discussed the major causes of error inherent in the use of a practical slotted line. These errors, which are systematic, result from the slotted line failing to adequately satisfy the design considerations listed above, and will be discussed in the following sections. Random errors, caused for example by frequency and level instabilities, have their origin in the circuitry external to the slotted line, and as such are amenable to improvements in circuit components and techniques.

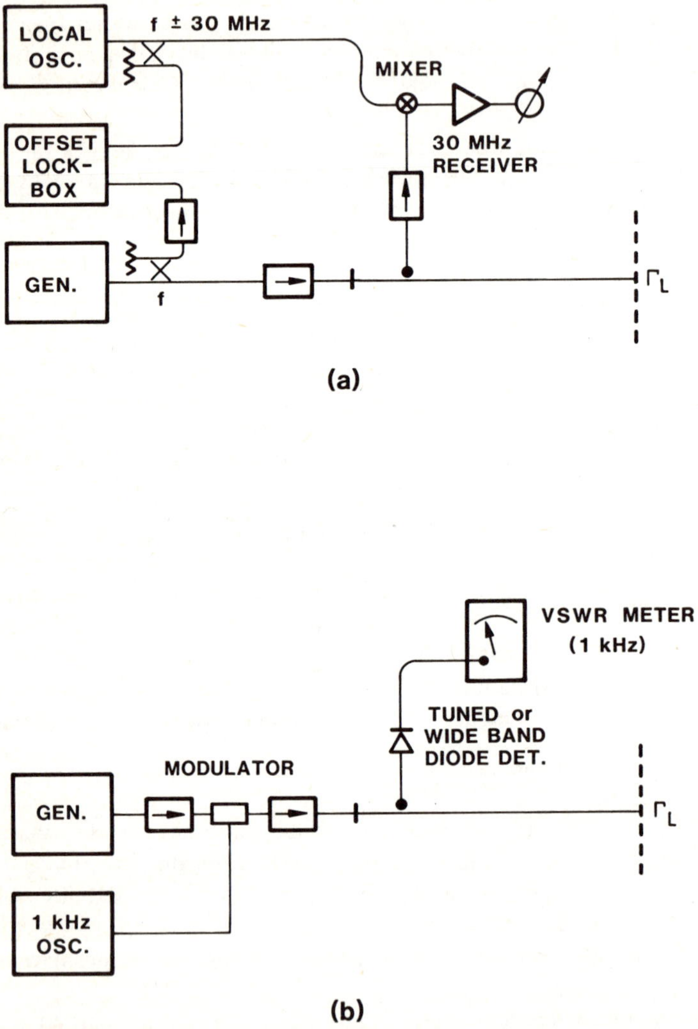

Fig. 4.3. *Typical circuits using a movable probe to detect the voltage or current distribution along a transmission line*
a Linear detection with a heterodyne system
b Square-law detection using a tuned or wideband diode detector operated in its square-law region

4.2.1.1 Loading by the probe: When the effect of the probe on the transmission line may be represented by a normalized admittance Y'_P across the line [6], the flow graph of the circuit is as shown in Fig. 4.6*a*, where it is assumed that the generator has been matched to the slotted line. The reflection coefficient Γ_P of the probe is related to Y'_P by

$$\Gamma_P = -Y'_P/(2+Y'_P) \tag{4.6}$$

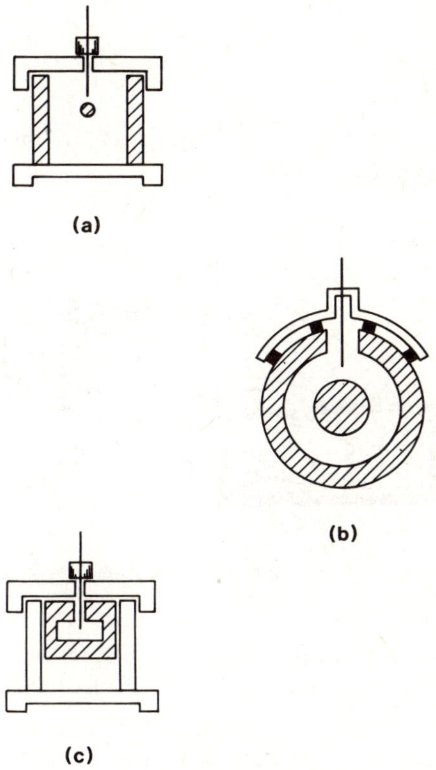

Fig. 4.4. *Some forms of single probe standing wave detectors*
a Slab line
b Coaxial slotted line
c Waveguide slotted line

Figs. 4.6*b* and *c* show the relevant flow graphs for the calculation of the incident and reflected fields at the probe. Using the flow graph reduction rules from Chapter 2 we obtain

$$V = V_{inc} + V_{ref} = \frac{(1+\Gamma_P)(1+\Gamma_L e^{-j2\phi})}{1-\Gamma_P\Gamma_L e^{-j2\phi}} \tag{4.7}$$

In practice, a tuner is often associated with a slotted line probe, and is adjusted to maximize the detected signal. Then Y'_P is real, and from eqn. 4.6 $\Gamma_P = -\rho$ where ρ is real and positive. An examination of eqn. 4.7 under these conditions reveals that the locations of the maxima and minima of $|V|$ are unchanged, but the voltage standing wave ratio S_m measured with such a probe is related to the VSWR of eqn. 4.3 by

$$S = S'S_m \tag{4.8}$$

where

$$S' = \frac{1+\rho|\Gamma_L|}{1-\rho|\Gamma_L|} \tag{4.9}$$

As a numerical example, when $S = 10$, and $\rho = 0.01$, then $S' = 1.016$ and the measured VSWR would be $S_m = 9.84$.

Fig. 4.5. *Current distributions in the walls of a rectangular waveguide operated in the fundamental TE_{10} mode*

In general, with the generator mismatched with reflection coefficient Γ_G, the locations of the maxima and minima remain unchanged, but S is given by eqn. 4.8 with

$$\left[\frac{(1+\rho|\Gamma_L|)^2+(1+|\Gamma_G||\Gamma_L|)^2-1}{(1-\rho|\Gamma_L|)^2+(1+|\Gamma_G||\Gamma_L|)^2-1}\right]^{1/2} \leqslant S' \leqslant$$

$$\left[\frac{(1+\rho|\Gamma_L|)^2+(1-|\Gamma_G||\Gamma_L|)^2-1}{(1-\rho|\Gamma_L|)^2+(1-|\Gamma_G||\Gamma_L|)^2-1}\right]^{1/2} \tag{4.10}$$

subject to the condition $\rho|\Gamma_G| \ll 1$. Using the same numerical example as above, but introducing a generator mismatch of $|\Gamma_G| = 0.05$ gives $1.015 \leqslant S' \leqslant 1.018$.

It can be seen from eqns. 4.9 and 4.10 that the effect of probe loading is small for small VSWR, but becomes increasingly significant as the VSWR increases. This is because the probe loading has greater effect on the larger voltage maxima, where the impedance is higher. Tomiyasu [7] has suggested a systematic method of probe withdrawal for measuring a high VSWR when the probe loading is significant. After a minimum has been located, the probe is shifted away until the detected level increases by (say) 10 dB. The probe carriage is then locked, and the probe withdrawn until the detected level is reduced to that at the minimum. This process is continued until the maximum is reached, at which time the VSWR can be calculated as the sum of all the 10 dB steps plus the final indication.

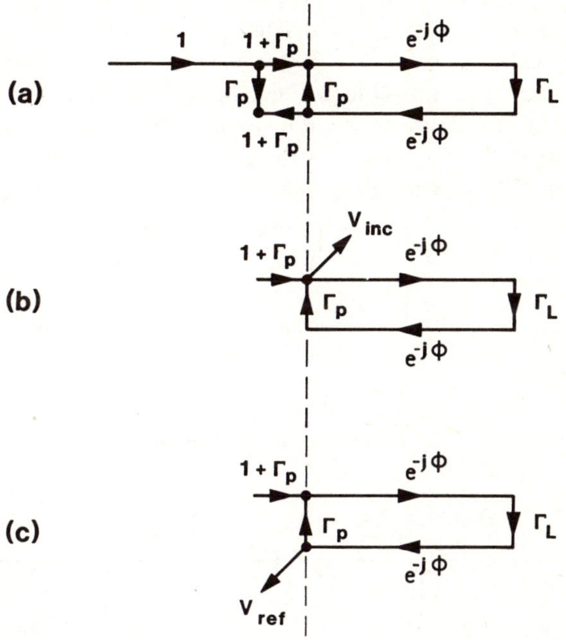

Fig. 4.6. *Transmission line loaded by a probe represented as a shunt admittance*
a Flow graph of complete circuit
b Flow graph reduced to find the incident voltage
c Flow graph reduced to find the reflected voltage

The high sensitivity associated with heterodyne detection allows the probe to be withdrawn until its effect is negligible. However, if heterodyne detection is not available and probe loading is suspected, it may be measured by introducing a second probe into the transmission line between the first probe and the generator. The slotted line is terminated in a matched load, and the probe under test is moved along the line. Provided the probe loading is not too great, ρ may be approximated from the maximum and minimum voltages detected by the second probe, as

$$\rho = \frac{|V_{max}| - |V_{min}|}{|V_{max}| + |V_{min}|} \qquad (4.11)$$

It is possible to perform this measurement in an imperfectly matched system because the probe reflection ρ becomes identifiable from the sum of all other reflections when the probe is moved. The error due to changes in multiple reflections will be of second order. The value obtained for ρ may then be used with eqns. 4.9 or 4.10 to correct subsequent measurements.

Alternatively, the admittance representing the probe loading may be measured by a resonance method [8]. The slotted line is terminated in almost full reflections incorporating loose coupling to the generator and detector, as shown for coaxial line in Fig. 4.7 [9]. The loose coupling can be achieved in waveguides by using iris sections. The generator frequency is adjusted near the frequency of interest until the slotted line resonates as an $n\lambda_1/4$ resonator with the probe positioned at a voltage minimum, and the Q-factor Q_1 measured by determining the 3 dB bandwidth. In order to ensure that the coupling is sufficiently loose so as not to affect the measurement, it should be reduced until no change is observed in the measured Q-factor. Then, with the probe located at a voltage maximum, the frequency is reduced until the line resonates now as an $n\lambda_2/4$ resonator with a Q-factor Q_2.

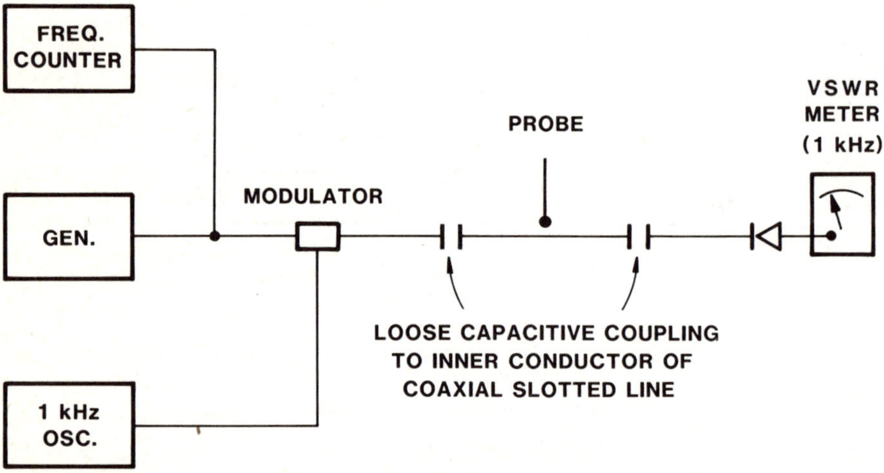

Fig. 4.7. *Circuit to measure the probe loading of a standing wave detector by a resonance method*

The normalized shunt susceptance B'_P of the probe is given by

$$B'_P \simeq \frac{n\pi}{2}\left(1 - \frac{\lambda_1}{\lambda_2}\right) \tag{4.12}$$

and, if the slotted line is assumed lossless, the normalized shunt admittance G'_P of the probe is [10]

$$G'_P \simeq n\pi/4Q_P \tag{4.13}$$

where the probe Q-factor Q_P is

$$Q_P = Q_1Q_2/(Q_1 - Q_2) \tag{4.14}$$

Typical results obtained on a 50 ohm commercial slotted line with the probe tuned for maximum sensitivity near 2 GHz were $n = 18$ at $f_1 = 1912.41$ MHz with $Q_1 = 1250$, and $f_2 = 1912.21$ MHz, $Q_2 = 1111$.* From eqns. 4.12 and 4.13 the probe capacitance and admittance are 4.9 fF and 28 μS, respectively, and from eqn. 4.6, $|\Gamma_P| = 0.0016$, which demonstrates the sensitivity of the method. Although the probe was tuned, the apparent probe capacitance is non-zero because of carriage capacitance.

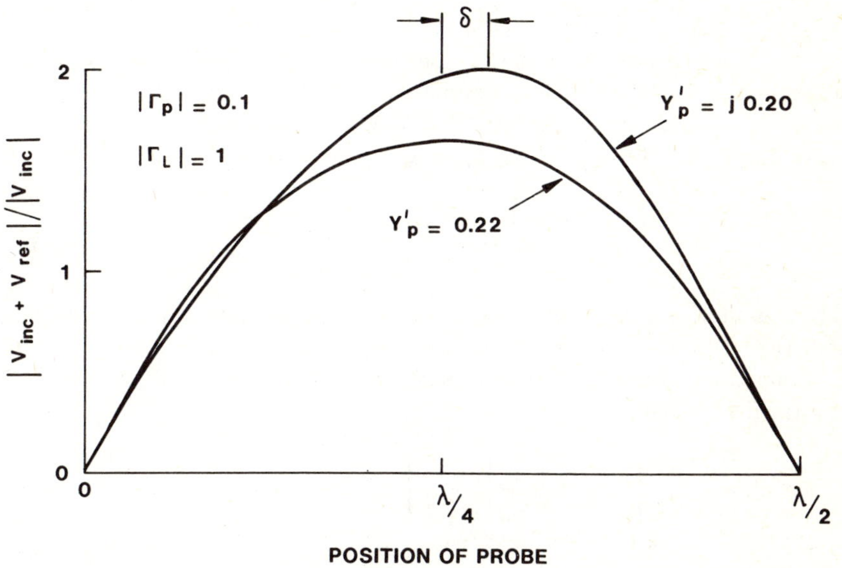

Fig. 4.8. *Distorting effect of a probe, having pure real or pure imaginary shunt admittance, on the voltage standing wave pattern. Real loading reduces the voltage peak, and imaginary loading shifts the peak*

Fig. 4.8 shows the effect on the line voltage of both a real and an imaginary normalized probe admittance Y'_P, with the generator matched and the line terminated in a short circuit. $|\Gamma_P|$ has been given the high value of 0.1 to emphasize its effect in the drawing. When Y'_P is imaginary, it can be seen that although the magnitude of the line voltage is unchanged, the voltage pattern is asymmetrical. The shift of the maximum from the midway point between the minima is

$$\delta = \sin^{-1}\left[\frac{2|Y'_P||\Gamma_L|}{\{4 + |Y'_P|^2[1 + |\Gamma_L|^2]^2\}^{1/2}}\right] \tag{4.15}$$

* P.I. Somlo

However, since the determination of the phase ϕ_L is usually done by considering the shift in the minima when a short circuit and the unknown are connected successively to the line, the error introduced into the measurement of ϕ_L by reactive probe loading is

$$\delta_e = \frac{|Y'_P|}{4}(1-|\Gamma_L|)^2, \quad |Y'_P| \ll 1 \tag{4.16}$$

For example, if $|\Gamma_L| = 0.1$ and $Y'_P = j0.02$ ($|\Gamma_P| \simeq 0.01$) the error introduced into the measurement of ϕ_L is 0.23 degrees. It can be concluded that with the generator matched, slotted line measurements are quite tolerant of reactive probe loading.

4.2.1.2 Line losses: In order to obtain maximum accuracy in slotted line measurements, the effect of line loss must be taken into account. For the range of parameters encountered in practice, the VSWR S which would be measured on a lossless line is related to the VSWR S_m measured on a lossy line by

$$S = \coth\left[\tanh^{-1}\left\{\frac{1}{S_m} \pm \frac{\alpha\lambda_g}{4}\right\} - \alpha l\right] \tag{4.17}$$

where l is the distance from the termination to the maximum measured, and the upper (lower) sign is chosen if the minimum measured is closer to (further from) the termination than the maximum. If S_m is not very large, eqn. 4.17 reduces to the form given by Ginzton [11]

$$S = \coth\left[\tanh^{-1}\left\{\frac{1}{S_m}\right\} - \alpha l\right] \tag{4.18}$$

ϕ_L is obtained using eqn. 4.5 as before. An alternative calculation procedure is to evaluate the effective reflection coefficient Γ_m at the plane of the measured minimum, and then determine Γ_L using

$$\Gamma_L = \Gamma_m e^{2\gamma l} \tag{4.19}$$

For low-loss slotted lines (attenuation per guide wavelength less than 0.01 dB) the value of α may either be determined theoretically from the conductor parameters, or may be measured by one of a number of methods, including direct measurement, or use of a resonance method as described in Section 4.2.1.1, where α is calculable from $\alpha = \beta/2Q$. Alternatively, α may be determined by connecting a well-matched load to the measuring port of the slotted line, recording the probe voltage as the probe is moved, and extracting the average slope from the recording. If the probe voltages at two successive minima and the intervening maximum are noted when the slotted line termination provides a full reflection, α may be calculated from

$$\alpha = \frac{2}{\lambda_g}\left[\frac{|V|_{\min 2} - |V|_{\min 1}}{|V|_{\max}}\right] \text{ nepers/m} \tag{4.20}$$

which is presented in nomograph form in Fig. 4.9 [12].

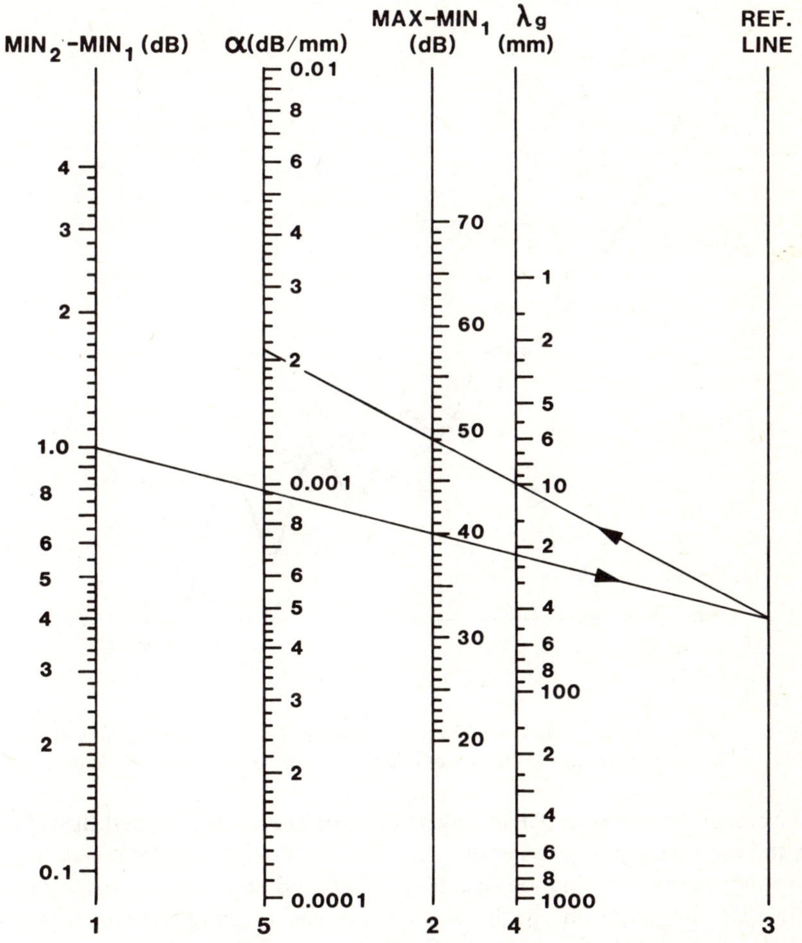

Fig. 4.9. *Nomograph to determine transmission line attenuation from measured values of two voltage standing wave minima and the intervening maximum*

If the attenuation constant is large, the voltage along a slotted line terminated in an open circuit may appear as shown in Fig. 4.10. Sufficiently far from the termination (to the left of position A in Fig. 4.10) local minima cease to exist, and therefore conventional VSWR measurements become impossible to perform.

Gardiol [13] suggests finding the relevant parameters by starting from an initial estimate and using a computer optimization procedure.

4.2.1.3. Line non-uniformities: Practical slotted lines are electrically non-uniform because of the existence of the slot ends, and dimensional variations, which result in a variation of the characteristic impedance or probe penetration with position.

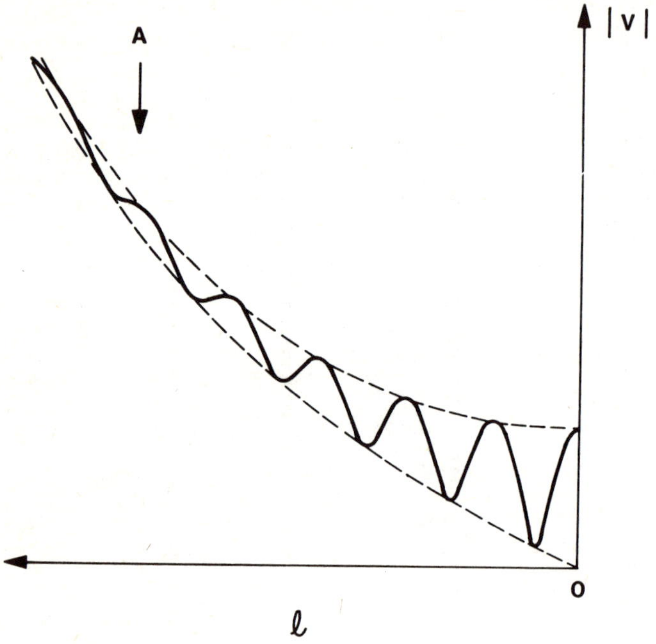

Fig. 4.10. *Voltage standing wave pattern along a transmission line having high attentuation constant. Note that to the left of position A, voltage minima cease to exist*

The non-uniformities which are independent of probe position, and the reflection from the measuring port connector, may be taken into account by representing the slotted line as the combination of a perfect line and an error box, as described in Section 2.4. In particular, if the slotted line can be considered lossless and of sufficient quality that $|S_{11}| < 0.01$, $|S_{21}|^2 \simeq 1$, then for the measurement of small reflection coefficients eqn. 2.26 may be approximated by

$$\Gamma_L \simeq \Gamma - S_{11} \tag{4.21}$$

The effect of the variation of probe penetration with position along the line may be taken into account by first determining the variation, and then correcting measured results by the appropriate factor. For coaxial slotted lines, the variation of probe penetration may be determined directly by examining the probe output as its position is varied while the line is driven by a comparatively high level signal at an audio frequency, or alternatively, by examining the probe response to a low-frequency signal applied to the line during the intervals when a square-wave modulated microwave signal is off [14].

4.2.1.4 Detector law: The output V_{dc} of a slotted line detection system can be related to the magnitude V_{rf} of the standing wave at the probe by

$$V_{dc} = k\, V_{rf}^{\eta} \tag{4.22}$$

where k is a constant of sensitivity, and the exponent η is, in general a function of V_{rf}.

With a superheterodyne detection system, it is not difficult to ensure that η has a constant value of unity, resulting in a linear detector law which introduces no errors into the measurement.

The simpler and more common diode detection systems are often used under the assumption (either explicit, or implied on the scale of a VSWR meter) that $\eta = 2$ for perfect square-law detection. For accurate measurements it is important to check the properties of a nominally square-law detector, as can be seen from the curves in Fig. 4.11 which show the measured characteristics of a broadband coaxial diode detector at 5 GHz. There is a slight curvature in the curve of microwave power against detector output, and a pronounced deviation of the exponent η from the constant value of two, which is seen to be valid only at very low signal levels.

In order to estimate the effect of a nominally square-law detector being accepted as perfect, we can approximate the values of exponent which would be encountered in a VSWR measurement by an average value η. Then, from eqn. 4.2, the detector output is

$$V_{dc} = k\,[1 + |\Gamma_L|^2 + 2|\Gamma_L| \cos(2\beta l - \phi_L)]^{\eta/2} \tag{4.23}$$

If the detector is incorrectly assumed to be square-law, the measured VSWR S_m is related to the true VSWR S by

$$S_m = S^{\eta/2} \tag{4.24}$$

and for $\eta < 2$, the value $|\Gamma_m|$ deduced from the measurement is less than the true value $|\Gamma_L|$. The amount of this error is illustrated in Fig. 4.12.

If it is impractical to operate a nominally square-law detector in the region where $\eta = 2$, then a high resolution variable attenuator can be placed in the signal line, and the VSWR deduced from the attenuator readings necessary to maintain the detector output at a constant level. Alternatively, the detector characteristics can be obtained over the dynamic range to be used, and the detector output corrected for these characteristics.

Another method is to use the slotted line diode detector output as a levelling signal, thereby maintaining constant the microwave level at the probe. The response of a linear microwave power meter placed inside the levelling loop (where sufficient power is available for its satisfactory operation) will be the true VSWR as the probe carriage is moved. This method is limited to VSWR values which can be accommodated by the dynamic range of the power meter.

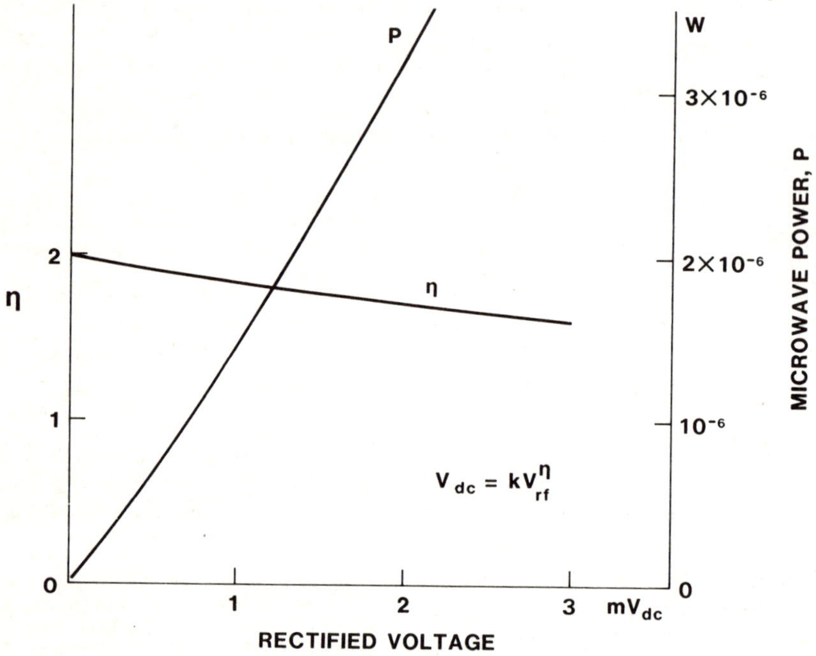

Fig. 4.11. *Measured rectifying characteristics of a microwave diode, showing the deviation from square-law behaviour*

4.2.1.5 Flow graph of a practical slotted line: The flow graph of a practical slotted line is shown in Fig. 4.13. Γ_G, Γ_P, and Γ_L are the reflection coefficients of the generator, probe, and unknown, respectively, and $\boldsymbol{S}$ is the S parameter matrix of the error box which accounts for the combined residual reflections. l_1 and l_2 are the lengths of line of propagation constant γ on each side of the probe, and as the probe is moved their sum remains constant. The function $f(l)$ describes the variation of probe coupling with position along the line, and the variation of detector output with input level is described by $g(V)$. The slotted line may be analysed by using a flow graph reduction technique as given in Section 2.3.

4.2.2 Special techniques using slotted lines

4.2.2.1 Measurement of large standing wave ratios – Roberts and von Hippel method: As has already been mentioned, the direct measurement of a large VSWR using a slotted line is difficult becaue the near-zero signal at the location of the minima can disappear in the noise of the detector system, and because the large difference in signal levels between the maxima and minima places excessive demands on the knowledge of the detector law.

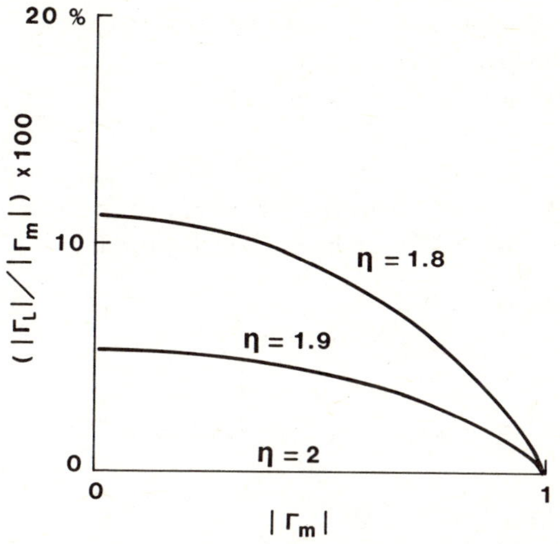

Fig. 4.12. *Error in the reflection coefficient magnitude if square-law behaviour is assumed in measuring a voltage standing wave pattern. η is the detector law exponent*

These limitations may be overcome by measuring the magnitude of the standing wave voltage only near a minimum, where the impedance is very low (Z_0/S). This enables the probe penetration, and consequently the detected signal, to be increased without introducing significant loading. The procedure is to note the separation w_1 between the two points at which the standing wave voltage has a convenient magnitude of $|V_1|$ on either side of the minimum, and the separation w between the two points at which the voltage has magnitude $K|V_1|$ on either side of the minimum. Then S may be calculated using [15]

$$S = \left[\frac{K^2 - \cos^2\left(\dfrac{\pi w}{\lambda_g}\right)}{\sin^2\left(\dfrac{\pi w}{\lambda_g}\right) - K^2 \sin^2\left(\dfrac{\pi w_1}{\lambda_g}\right)} \right]^{1/2} \tag{4.25}$$

In order to obtain accurate results using this method, the carriage position should be sensed accurately by using, for example, a dial gauge, micrometer drive, or laser ranging system, and the line loss should also be taken into account. The value chosen for K should not be so large that the increasing probe penetration, which is acceptable near the minimum, causes line loading at the measurement points.

If the voltage on the slotted line, and the probe penetration is such that $|V_{min}|$

can be measured, we can set $|V_1| = |V_{min}|$, $w_1 = 0$ in eqn. 4.25 and obtain the Roberts and von Hippel formula [16, 17]

$$S = \frac{\left[K^2 - \cos^2\left(\dfrac{\pi w}{\lambda_g}\right)\right]^{1/2}}{\sin\left(\dfrac{\pi w}{\lambda_g}\right)} \tag{4.26}$$

When a square-law detector is used, it is convenient to choose $K = \sqrt{2}$, which corresponds to a doubling of the detector output at the points separated by w compared to the output at the minimum. Then, provided $\pi w/\lambda_g \ll 1$, S may be approximated from eqn. 4.26 by

$$S \simeq \frac{\lambda_g}{\pi w} \tag{4.27}$$

The error associated with this approximation is shown in Fig. 4.14.

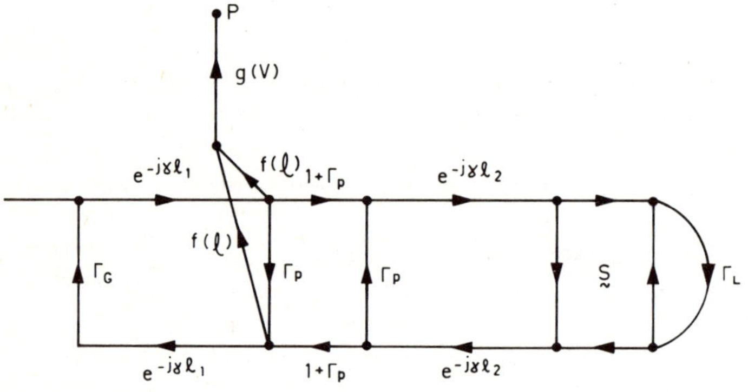

Fig. 4.13. *Complete flow graph of a practical slotted line circuit*
$f(\ell)$ describes the variation of probe coupling with position, and $g(V)$ represents the detector law. The length of the slotted line is $\ell_1 + \ell_2$

4.2.2.2 Measurement of small standing wave ratios: The accurate measurement of small reflections using slotted lines is hindered by the presence of mechanical imperfections which cause the probe penetration to vary and the probe to deviate from its intended path. Other mechanical and electrical imperfections create residual reflections. Some of the techniques for overcoming these problems are discussed in detail by MacKenzie [18].

A computer assisted method developed by Harvey [19] partially compensates for the mechanical imperfections of the line by recording digitally at least two periods of the standing wave pattern for each of two offset (nominally) matched loads, a phase reference short circuit, and the unknown. Each of the patterns is then divided into eight segments of length $\lambda/2$ but offset by $\lambda/16$ from each other, as shown in Fig. 4.15, which is the output from a square-law detector. It follows from eqn. 4.4 that if a square-law detector is used, Fourier analysis of the standing wave over a segment will result in a fundamental component related to Γ_L.

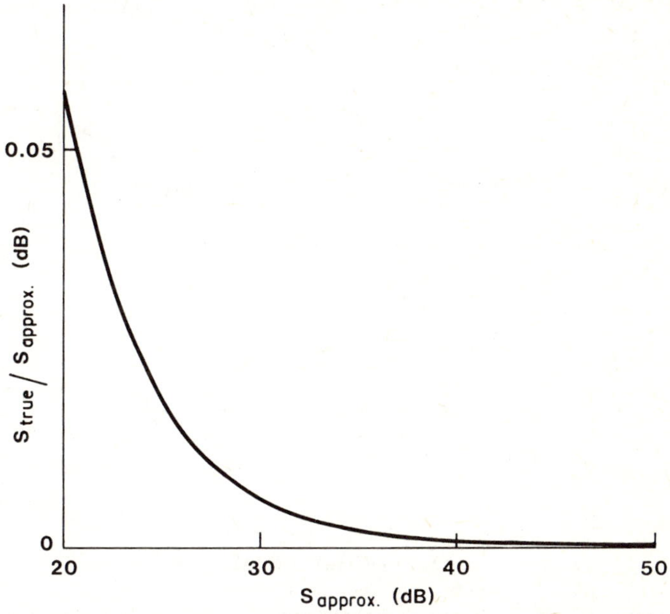

Fig. 4.14. *Error resulting from the use of the approximate formula in eqn. 4.27 when measuring a large standing wave ratio using the Roberts and von Hippel method* [*16*]

For each segment, the principal element S_{11} of the error box is approximated from the fundamental Fourier components ${}_1\Gamma_m$, ${}_2\Gamma_m$ derived when the two offset (nominally) matched loads are connected. If the offset is $\lambda/4$, then it follows from eqn. 4.21 that

$$S_{11} \simeq ({}_1\Gamma_m + {}_2\Gamma_m)/2 \qquad (4.28)$$

and this value of S_{11} is subtracted from all other results obtained over this segment range.

The eight estimates of the unknown obtained from analysis over each of the eight segments are averaged to obtain the best estimate of the unknown. The advantage of this procedure is that since each estimate is computed using only the same section of line, mechanical imperfections will be common and likely to affect the fundamental Fourier component in a similar fashion. The final averaging process partially compensates for these imperfections and for noise. The relative error in the value of Γ_L obtained using the fundamental Fourier component when the detector is not square-law is shown in Fig. 4.16.

4.2.2.3 Tuning out residual reflections: When a small reflection coefficient is being measured using a slotted line, it is helpful to interpose a tuner between the measuring port and the unknown and adjust it until the residual reflection seen by the probe is minimized. This procedure is valid if all residual reflections are sufficiently small that multiple reflections can be neglected, and therefore the residual reflections may be superimposed so that they appear to originate from one discontinuity.

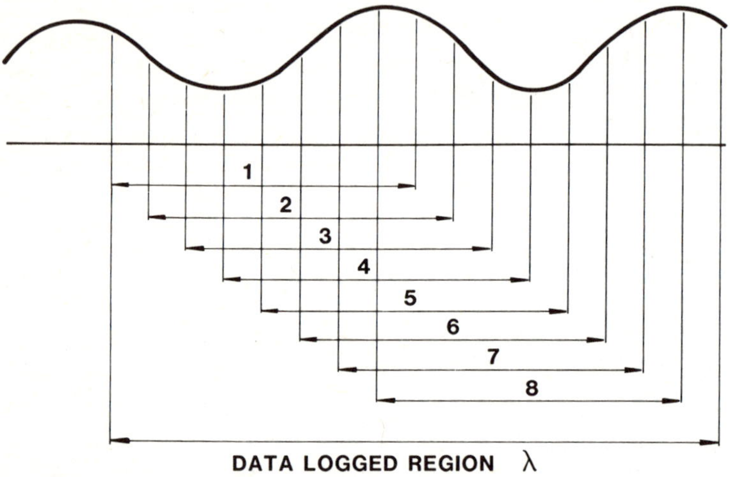

Fig. 4.15. *Segments used by Harvey in Fourier analysing the output of a square-law standing wave detector* [*19*]

If a perfectly matched load was attached to the tuner, and the probe moved along the slotted line, the tuner could be adjusted until the variation in the signal detected by the probe was minimized. In practice, a perfectly matched load does not exist, but a procedure described by Weinschel *et al.* [20] enables the tuner to be adjusted with respect to the characteristic impedance of a length of precision airline. A precision airline containing a sliding load is connected to a tuner, which is in turn connected to the measuring port of a slotted line, as shown in Fig. 4.17. The sliding load is mechanically linked to the probe carriage. If the residual reflections are sufficiently small that multiple reflections can be neglected, their effect may be represented by a single reflection which is independent of the position of the (imperfect) load. Therefore, as the load and attached carriage are moved, the probe will detect no variation in the reflection from the load, but only the variation of the residual reflection along the slotted line. The tuner may then be adjusted to minimize this variation. When this procedure is completed, the precision airline can be removed, leaving the slotted line with tuner ready for use.

The flow graph representation of the circuit in Fig. 4.17 is shown in Fig. 4.18 with the generator mismatched. In order to simplify the analysis of this tuning procedure, the effect of loading by the probe is neglected. The slotted line length $l_1 + l_2$ remains constant, as does the probe to load separation $l_2 + l_3$. Γ_r is associated with the small residual reflections referred to the output port of the slotted line, and Γ_t and θ are associated with the two-port representation of a tuner (Fig. 3.13*a*). Flow graph reduction techniques show that after adjustment of the tuner and removal of the sliding load, there remains a residual reflection which affects the measurement of an unknown. The VSWR of the residual is

$$S = \frac{1 + |\Gamma_L||\Gamma_G|}{1 - |\Gamma_L||\Gamma_G|} \tag{4.29}$$

and therefore for the procedure to result in the complete cancellation of residual reflections, the generator must be matched. For example, to reduce the residual VSWR below 1·001 when the magnitude of the reflection coefficient $|\Gamma_L|$ of the sliding load is 0·02, the magnitude of the generator reflection coefficient $|\Gamma_G|$ must be less than 0·025.

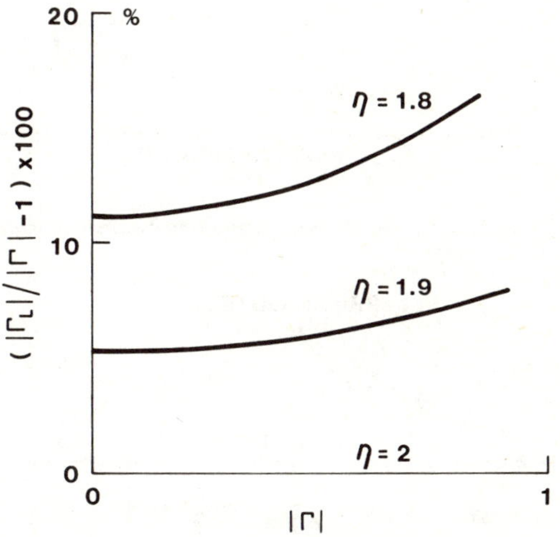

Fig. 4.16. *Error in the reflection coefficient magnitude if square-law detector behaviour is assumed when Fourier analysing the detector output.* η *is the detector law exponent*

4.2.2.4 Measurement when load power must be low: When the power impinging on the unknown must be kept low, the signal may be introduced across the probe, and the slotted line terminated in the unknown at one end, and the (not necessarily matched) detector at the other [11]. The requirement for low power into the unknown occurs, for example, when measuring the impedance of a diode at low levels. If the unknown is non-linear, care must be taken to ensure that the harmonics generated are absorbed before reaching the slotted line detector, and are not reflected back to the unknown.

4.2.2.5 Error correction by modelling: Clifton [21] describes a computer-corrected slotted line measuring method in which a multisection transmission line model is constructed to represent the measuring set-up. The slotted line effective resistivity and the contact resistance at the measuring connector are determined as functions of frequency. Therefore, provided the model is sufficiently detailed, the method has the advantage that frequency dependence is included, and measurements are corrected at any frequency for which the model is valid. The particular model proposed by Clifton is appropriate when the signal is introduced across the probe, as described in Section 4.2.2.4.

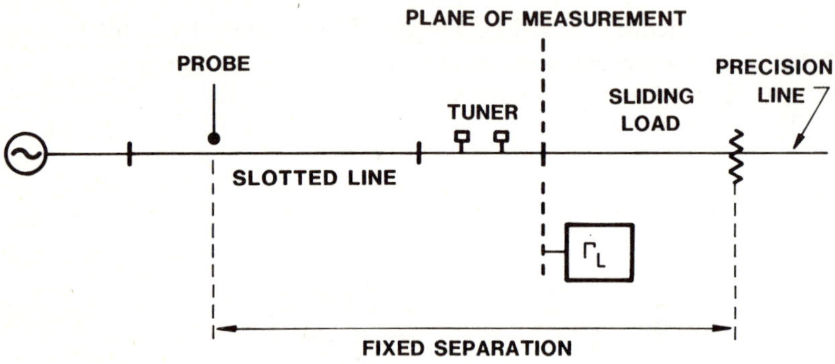

Fig. 4.17. *Weinschel's method for tuning out slotted line residual reflections* [20]

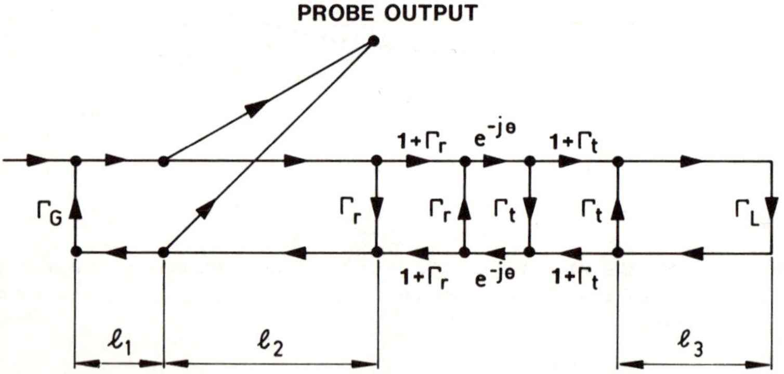

Fig. 4.18. *Flow graph representation of Weinschel's circuit to tune out slotted line residual reflections* [20]

4.2.2.6 Swept frequency measurements: If knowledge of the VSWR of a device is required at a large number of frequencies, it is possible to use the slotted line in a swept frequency mode.

A schematic of a circuit suitable for the method described by Sorger and

Weinschel [22], Adam [23], and Sorger [24] is shown in Fig. 4.19. Two slotted lines having detectors and probes with matched frequency responses are connected together. The first probe is used as a reference detector whose output is held constant by the levelling loop, thereby compensating for the frequency response of the measuring probe. The *X*-deflection of an *X-Y* recorder or storage oscilloscope is proportional to frequency, and the *Y*-deflection records the measuring detector output, which it is convenient to plot logarithmically. If the measuring probe is moved asynchronously along the slotted line while the frequency is swept over the range of interest, a succession of traces will be recorded. The maximum and minimum values of the envelope of these traces provide, at each frequency, the VSWR of the unknown, and may be conveniently measured in decibels. To increase the ease with which the traces may be interpreted, they can be normalized to a common baseline [14]. The calibration of the envelope is obtained from the envelope of a reference reflection.

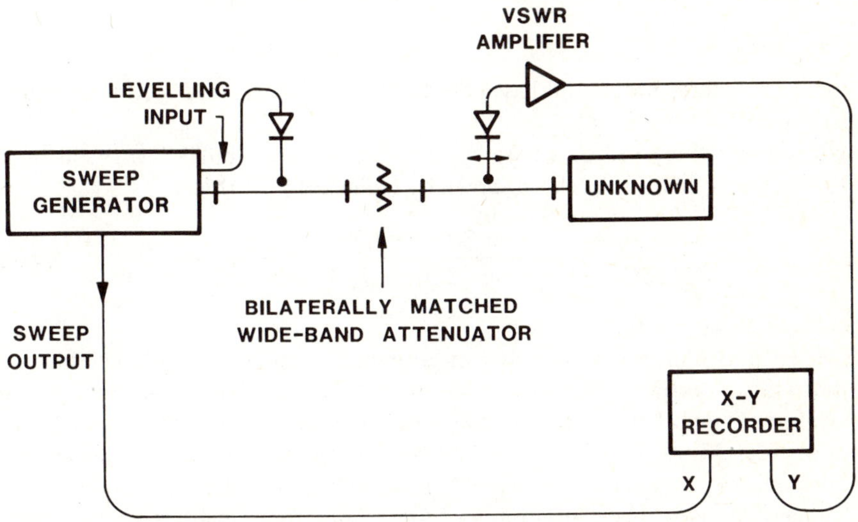

Fig. 4.19. *Swept-frequency VSWR measurement circuit with compensation for probe frequency response after Sorger and Weinschel* [22] *and Adam* [23]

This method provides only the VSWR of the unknown, and therefore only the magnitude of its reflection coefficient. The magnitude of the residual reflection coefficient, which should be less than the magnitude of the reflection coefficient of the unknown, can be determined using the coupled sliding-load technique described in Section 4.2.2.3.

4.2.3 Methods without slotted lines

4.2.3.1 Fixed multiprobe methods: If three or more probes are fixed at appropriate intervals along a transmission line, it is possible to deduce from their outputs the impedance of the termination. Although many of the causes of error associated with slotted line measurements apply, along with the additional problem of uncertainty in the effective position of the probes, multiprobe methods have the advantage of no moving parts and are therefore more amenable to computerized operation.

Because of the fixed position of the probes, the usable bandwidth of a multiprobe device is limited. At low frequencies, the probes are quasi-paralleled, whereas as the frequency rises the outer probes become sufficiently close to $\lambda_g/2$ apart that they do not provide independent information.

It follows from eqn. 4.4 that if the nth probe is located a distance l_n from the termination, the output P_n from a square-law detector attached to it is

$$P_n \propto 1 + |\Gamma_L|^2 + 2|\Gamma_L| \cos(2\beta l_n - \phi_L) \tag{4.30}$$

Duffin [25] described a method using three probes $\lambda_g/8$ apart, with the first $\lambda_g/8$ from the termination. The normalized impedance of the load is then

$$|Z'_L|^2 = (P_1 + P_3 - P_2)/P_2, \quad X'_L = (P_1 - P_3)/2P_2 \tag{4.31}$$

A commercial three-probe unit based on the above equations has been described by Spector [26].

When using three probes whose spacings are arbitrary but known, and not a multiple of $\lambda_g/2$ apart, it has been shown by Caldecott [27] that

$$\Gamma_L \propto \frac{P_1(e_2 - e_3) + P_2(e_3 - e_1) + P_3(e_1 - e_2)}{\sin[2\beta(l_2 - l_3)] + \sin[2\beta(l_3 - l_1)] + \sin[2\beta(l_1 - l_2)]} \tag{4.32}$$

where $e_n = \exp(-j2\beta l_n)$.

Caldecott also suggested that the measurement confidence could be enhanced by introducing extra probes to provide extra (redundant) measures of Γ_L. This concept is somewhat akin to concepts associated with six-port methods (Section 7.2), but is not as general in that the separation of the probes must be known, and the probe reflections should be as small as possible.

Samuel [28] described a multiprobe system using four equally sensitive square-law probe detectors spaced $\lambda_g/8$ apart as shown in Fig. 4.20. If the two differences

$$\begin{aligned} P_1 - P_3 &\propto |\Gamma_L| \cos(2\beta l_1 - \phi_L) \\ P_4 - P_2 &\propto |\Gamma_L| \sin(2\beta l_1 - \phi_L) \end{aligned} \tag{4.33}$$

are applied to the X and Y plates of an oscilloscope, a Smith chart presentation is obtained. The orientation of the $\phi_L = \pi$ direction on the screen may be found by connecting a short circuit to the measuring port. Because the probe spacings are fixed, attempts to use this device over a 30% bandwidth may result in errors of up to 15% in $|\Gamma_L|$. It is interesting to note that $P_1 + P_3 \propto 1 + |\Gamma_L|^2$ and may be used as a levelling signal if $|\Gamma_L|$ is small [29].

A forerunner of the long-line reflectometer technique discussed in Section 5.4.3 was descibed by de Ronde [30] who used two square-law detectors a distance d

apart and separated from the unknown termination by a length l of transmission line, as shown in Fig. 4.21. The value of d is best chosen to be near one quarter of the mean value of the guide wavelengths at the extremes of the frequency sweep. When the generator frequency is swept, the difference between the two detector outputs is V_{diff} where

$$V_{\text{diff}} \propto |\Gamma_L| \sin\left(\frac{2\pi d}{\lambda_g}\right) \sin\left[\phi_L - (2l + d)\frac{2\pi}{\lambda_g}\right] \tag{4.34}$$

Provided l is sufficiently large, the oscillations caused by the changing electrical length of the line with λ_g are extremely rapid, and the envelope of V_{diff}, as seen on the screen of an oscilloscope, may be calibrated against that for a known termination, and hence provides $|\Gamma_L|$ as a function of frequency.

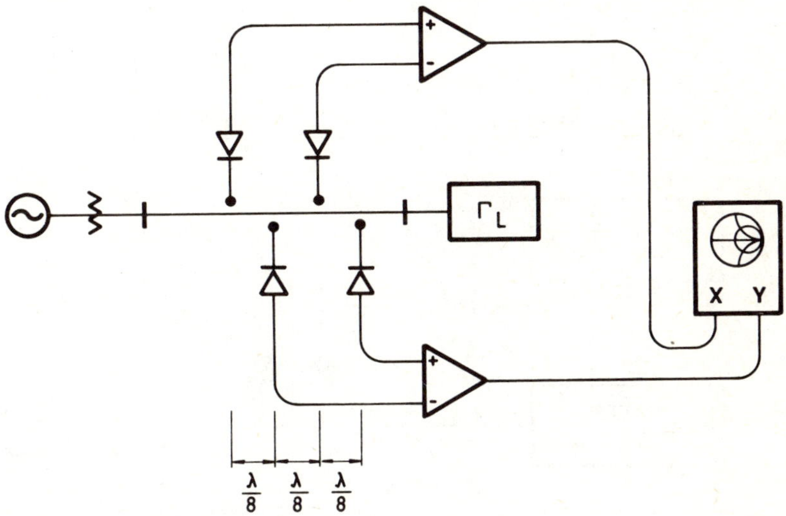

Fig. 4.20. *Samuel's fixed multiprobe system for obtaining a Smith chart display* [*28*]

Another multiprobe method, with couplers as the sampling elements, was described by Gottfried [31] who used the circuit in Fig. 4.22 to obtain a Smith chart display of reflection coefficient at millimetre wavelengths. The electrical lengths l_1 and l_2 are adjusted to be equal, and l_3 and l_4 to differ by $\lambda_g/8$. The output of the square-law detector C is proportional to $|\Gamma_L|^2$, and after processing, the components $|\Gamma_L| \cos\phi$ and $|\Gamma_L| \sin\phi$ can be extracted from the outputs of square-law detectors A, B and C, and applied to the X and Y inputs of a CRO to give a Smith chart display. With the position of the short circuits adjusted, the device is claimed to be capable of providing a real-time Smith chart display for reflection coefficient magnitudes greater than 0·005 over a 10 GHz swept bandwidth in the 50–75 GHz waveguide band. For small reflections the resolution is 0·001 in $|\Gamma_L|$, with an uncertainty over a 5 GHz swept bandwidth of 0·007 in $|\Gamma_L|$ and 4° in ϕ_L.

The circuit in Fig. 4.22, like that of the locating reflectometer discussed in Section 6.2.2.1 is a six-port network which provides a method of measuring Γ_L and obtaining a Smith chart display without resorting to the general theory of six-ports discussed in Section 7.2. A general six-port, for which post-detection digital processing based on six-port theory is used to determine Γ_L, may also be regarded as a multiprobe measurement method, and, like the automatic network analysers mentioned in Section 7.1, is capable of providing a Smith chart display. The amount of analogue processing necessary before such a display is available depends on the particular six-port configuration.

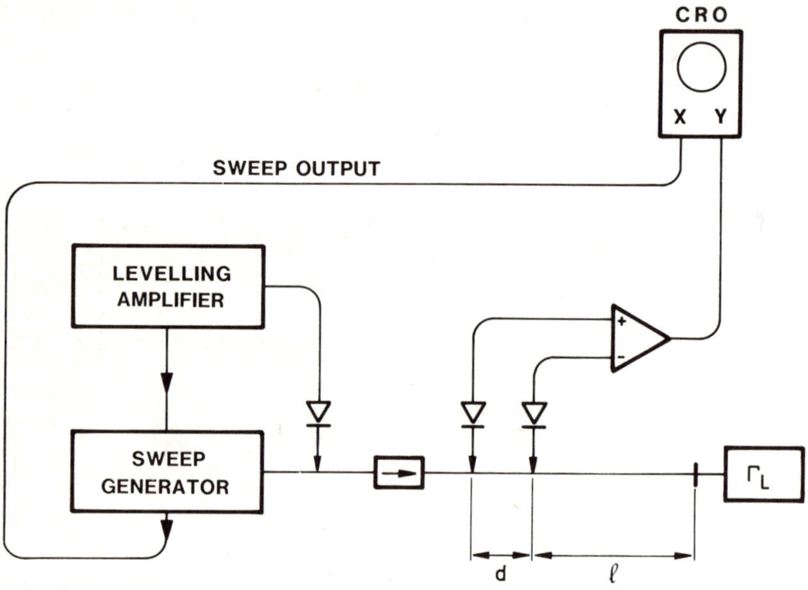

Fig. 4.21. *De Ronde's swept-frequency reflectometer* [*30*]

4.2.3.2 Coupler methods: In Fig. 4.23, the unknown is shown connected to port 2 of an ideal directional coupler of coupling coefficient c, with port 4 terminated in a sliding short circuit, and port 3 terminated in a matched detector [32]. The total signal incident on the detector is the vector sum of a signal V_s which has been reflected from the short circuit, and a signal V_L which has been reflected from the unknown, where, using Fig. 3.24b,

$$|V_s| = c(1-c^2)^{1/2}, \quad |V_L| = c(1-c^2)^{1/2}|\Gamma_L| \tag{4.35}$$

The phase of V_s relative to V_L may be varied by changing the position of the sliding short circuit, and the maximum and minimum signals at the detector become

$$|V|_{\max} = c(1-c^2)^{1/2}(1+|\Gamma_L|),$$
$$|V|_{\min} = c(1-c^2)^{1/2}(1-|\Gamma_L|) \quad (4.36)$$

The ratio of these signals is S as given by eqn. 4.3. The phase ϕ_L of the reflection coefficient may be found after first setting the sliding short circuit for a detector minimum when a short circuit is connected to port 2. Then if d is the extra distance by which the sliding short circuit must be moved away from port 4 for a detector minimum when the unknown is connected to the measuring port,

$$\phi_L = \pi\left[1 - \frac{4d}{\lambda_g}\right] \quad (4.37)$$

The advantage of coupler methods over slotted line measurements is the lower cost of components, particularly if cross-guide couplers are used for waveguide measurements [33]. The disadvantages are that the detector must be matched to prevent multiple reflections, and the source should be isolated or matched to prevent level changes in V_s with short-circuit position.

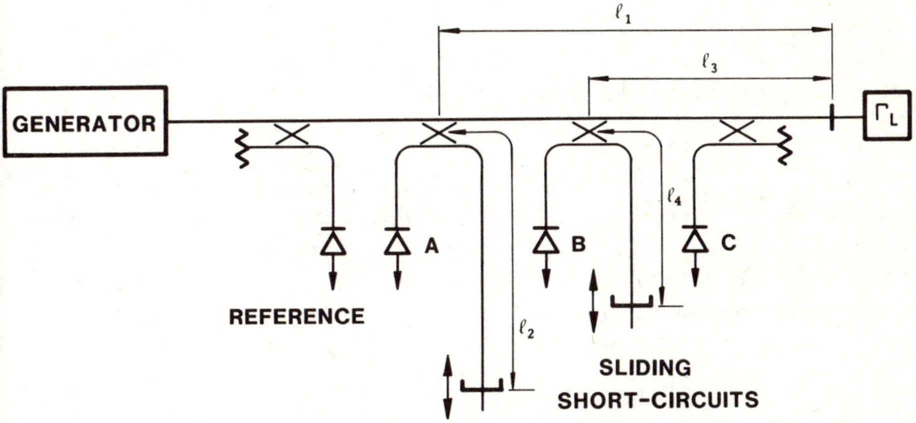

Fig. 4.22. *Gottfried's swept-frequency millimetre wave multiprobe reflectometer* [*31*]

The finite directivity of the coupler limits the minimum reflection coefficient which can be measured. For example, if the directivity of the coupler is 40 dB, the leakage signal at the detector is comparable with $|V_L|$ when $|\Gamma_L| = 0{\cdot}01$.

If an (ideal) attenuator is interposed between port 4 and the sliding short circuit, the apparent VSWR obtained with small $|\Gamma_L|$ may be increased, and that obtained with large $|\Gamma_L|$ may be reduced according to the formula

$$S_{app} = \left|\frac{2\alpha + |\Gamma_L|}{2\alpha - |\Gamma_L|}\right| \quad (4.38)$$

where α is the introduced attenuation. If the attenuator is imperfect, α cannot be taken directly from the attenuator dial, but must include the effect of reflections. When the attenuator setting and sliding short circuit are adjusted so that $|V|_{min} = 0$, the network may be regarded as a microwave bridge, and is discussed as such in Section 5.1.2.

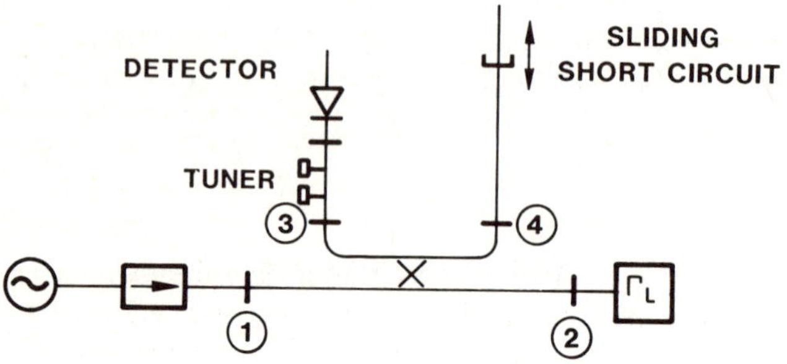

Fig. 4.23. *VSWR measurement circuit using directional coupler*

4.2.4 Measurement of generator impedance

Methods used to measure the output impedance of a signal generator may be categorized according to whether the measuring apparatus operates at the generator frequency (direct method) or at some other frequency (indirect method). For either category, the measuring apparatus must neither reflect nor inject sufficient signal into the generator to alter the output impedance.

4.2.4.1 Direct method: Fig. 4.24*a* shows the flow graph of the signals from a generator of reflection coefficient Γ_G connected by a length of lossless transmission line to a termination with reflection coefficient Γ_L. In this instance it is convenient to identify these as current reflection coefficients (the negative of voltage reflection coefficients as shown in Chapter 1), and let a represent the current in the termination due to the incident signal. Then

$$a = \frac{1}{1 - \Gamma_G \Gamma_L e^{-j2\phi}} \tag{4.39}$$

By definition, the current in the termination due to the reflected signal is $b = a\Gamma_L$, and therefore the total current in the termination is $a(1 + \Gamma_L)$. As the termination is slid along the transmission line, ϕ will vary, and the ratio of maximum to minimum current in the termination is

$$\left|\frac{I_{max}}{I_{min}}\right| = \frac{1 + |\Gamma_G||\Gamma_L|}{1 - |\Gamma_G||\Gamma_L|} \tag{4.40}$$

from which $|\Gamma_G|$ can be determined if $|\Gamma_L|$ is known.

The sensitivity of eqn. 4.40 to $|\Gamma_G|$ increases with $|\Gamma_L|$ and it is convenient to choose a sliding short circuit as the termination, as shown in Fig. 4.24*b*. If the source is affected by such a mismatch (some tightly coupled oscillators may even cease oscillating), a sliding load with $|\Gamma_L| < 1$ should be used.

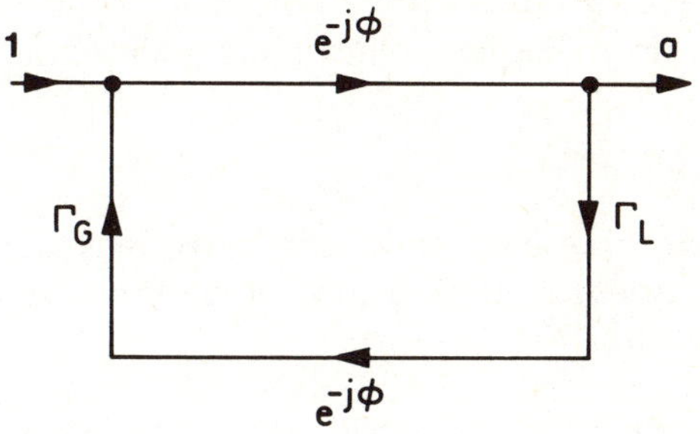

(a)

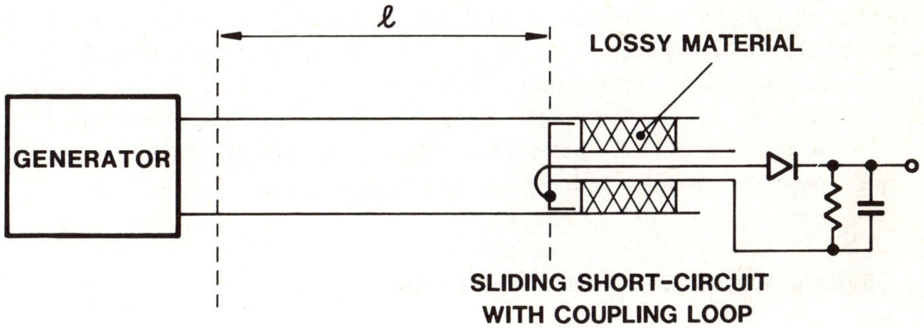

(b)

Fig. 4.24. *Direct method for the measurement of generator reflection coefficient*
a Flow graph representation of circuit
b Circuit showing sliding short circuit incorporating a coupling loop. The output of the detector provides the VSWR directly

A sliding short-circuit method with a coupling loop mounted in the termination was described for waveguides by Gilmer [34], and for coaxial lines by Somlo [35]. Weinschel [36] observed that similar results may be obtained by monitoring the voltage at an odd multiple of $\lambda/4$ from the short circuit, and using either a constant impedance line stretcher to produce the phase change or by coupling the sliding short circuit to the carriage of a slotted line. Alternatively, a directional coupler terminated in a sliding short circuit may be used to monitor either the forward or reflected signal. The VSWR of the generator is then the ratio of the maximum to minimum signal as the short circuit is slid.

The phase ϕ_G of the generator current reflection coefficient can be determined from the location of the sliding short circuit at a point of minimum signal. Then, from eqn. 4.39,

$$\phi_G = (2n+1)\pi + 2\phi \tag{4.41}$$

where $\phi = \beta l$ is the electrical angle between the phase reference plane and the short circuit at distance l. Another direct method of measuring generator impedance is mentioned in Section 5.3.1.

4.2.4.2 Indirect methods: In principle, many of the methods described in this book for the measurement of passive impedances can be used as indirect methods for the measurement of active impedances. However, the performance required of the detecting apparatus may be difficult to provide.

Because the generator output impedance may vary with frequency, it is desirable that the frequency of measurement be as close as possible to the generator frequency. The detection apparatus must have sufficiently narrow bandwidth to reject the generator signal. If there is no effective filtering of the generator signal before the detection apparatus, then the apparatus must have sufficient dynamic range to cope in the presence of this signal. For example, if there is no effective preselection at the input of a heterodyne detection system, the mixer is required to handle both the generator and measurement signals.

4.3 Methods using magnitude and phase detectors

4.3.1 Vector voltmeter method

The (complex) voltage V_1 measured across a transmission line at a distance l from the termination Z_L is the sum of the incident and reflected signals. The reflection coefficient Γ at the plane of measurement is the ratio of the reflected to incident signal and is given by

$$1 + \Gamma = \frac{V_1}{V_2} \tag{4.42}$$

where V_2 is the voltage measured when the line is terminated in its characteristic impedance Z_0. The reflection coefficient Γ_L of the unknown Z_L is related to Γ through eqn. 1.18, and is given by

$$\Gamma_L = \Gamma e^{2\gamma l} \tag{4.43}$$

The factor $e^{2\gamma l}$ may be found by calculating from eqn. 4.42 the reflection coefficient Γ_s at the measurement plane when a short circuit ($\Gamma_L = -1$) is the termination. Then from eqn. 4.43

$$\Gamma_s = -e^{-2\gamma l} \qquad (4.44)$$

Alternatively, Γ_L may be found using the Smith chart construction shown in Fig. 4.25. If a line of length $|1+\Gamma|$ is drawn from the point of zero impedance at an angle of $\arg(1+\Gamma)$ to the real impedance line, then Γ is represented by the closing vector from the centre of the chart. Γ_L is found by rotating the Γ vector anticlockwise through an angle $2\beta l$ about the centre of the chart, and extending the length by the factor $\exp(2\alpha l)$ in accordance with eqn. 4.43. This angle and factor may be found by plotting the vector $1+\Gamma_s$ on the chart in the same manner as described above. Then $2\beta l - 2n\pi$ is the angle by which the vector must be rotated anticlockwise until it lies on the real impedance axis near the zero impedance point, and from eqn. 4.44, $|\Gamma_s| = \exp(-2\alpha l)$.

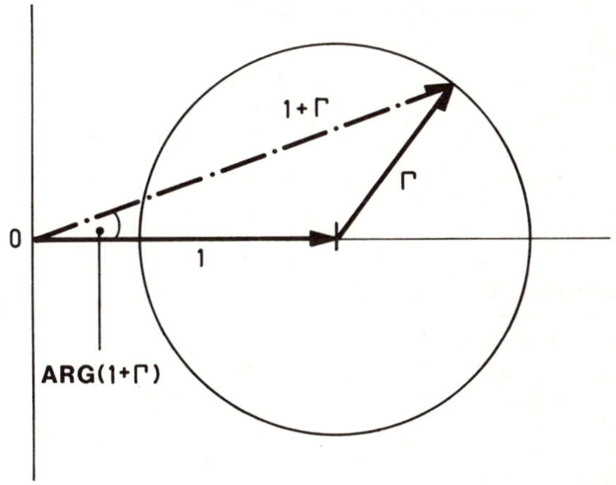

Fig. 4.25. *Smith chart construction to obtain* Γ *from* $|1+\Gamma|$ *and* $\arg(1+\Gamma)$*, as measured by the vector voltmeter method*

Commercially available vector voltmeters (VVM) provide a means by which the magnitude and the phase of the voltage on a coaxial line may be measured at frequencies up to about 1 GHz. The phase is measured relative to the phase detected by a second reference probe, but because only the ratio V_1/V_2 is needed for the above analysis, the location of this reference probe in the circuit is unimportant, at least in principle. Vector voltmeters obtain the phase information by down converting to an intermediate frequency where a direct phase comparison is performed. Because this down conversion may be accomplished by coherent sampling, the probes can inject a weak, but spectrally broad signal into the circuit. For such instruments, sufficient attenuation or isolation should be introduced between the probes in order that the operation of one does not interfere with the other.

Two impedance measuring circuits using a VVM are shown in Fig. 4.26. It is apparent that apart from the VVM, few components are needed. However, it can be seen from eqn. 4.42 that the measurement provides the quantity $1 + \Gamma$, and consequently, when $|\Gamma|$ is small the accuracy with which Γ can be obtained is

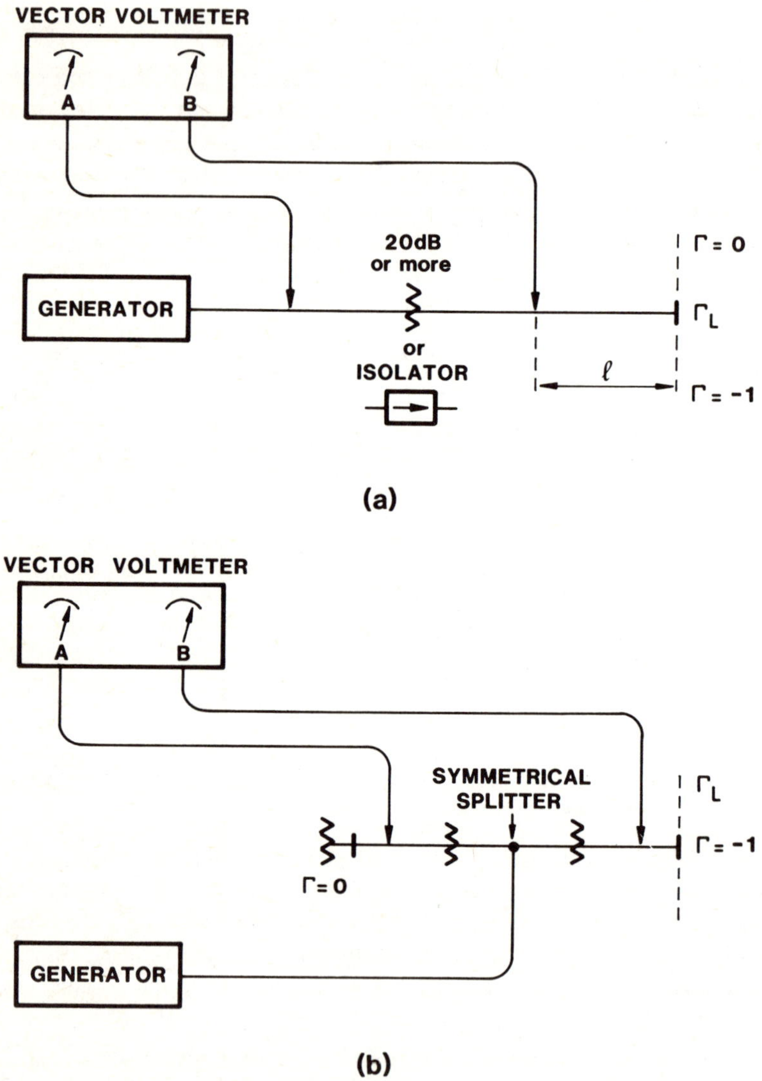

Fig. 4.26. *Impedance measurement with a vector voltmeter, but without a directional coupler*
a In-line configuration
b Split-line configuration. Instead of connecting a matched load to the measuring port, the amplitude reference is obtained by measuring the voltage at probe *A*

limited by the accuracy and resolution of the VVM. For example, for $\Gamma_L = 0{\cdot}1 \underline{/10^\circ}$, a $0{\cdot}1^\circ$ error in the argument of $1 + \Gamma$ can result in a 1° error in ϕ_L.

References

1 LECHER, E.: 'Study of resonance phenomena', *Annalen den Physik*, 1890, **41**, pp. 850–870

2 LABUS, J. W.: 'Measurement of resistive impedances at high frequencies', *Proc. IRE*, 1931, **19**, pp. 452–460

3 BARLOW, H. M. and CULLEN, A. L.: 'Micro-wave measurements' (Constable, London, 1950), p. 132

4 OLIVER, M. H.: 'Discontinuities in concentric-line impedance measuring apparatus', *Proc. IEE*, 1950, **97**, Pt. II., pp. 29–38

5 SORROWS, H. E., RYAN, W. E. and ELLENWOOD, R. C.: 'Evaluation of coaxial slotted-line impedance measurements', *Proc. IRE*, 1951, **39**, pp. 162–168

6 ALTAR, W., MARSHALL, F. B. and HUNTER, L. P.: 'Probe error in standing-wave detectors', *Proc. IRE*, 1946, **34**, pp. 33P–44P

7 TOMIYASU, K.: 'Decoupling probe method of measuring very high VSWR', *Proc. IEEE*, 1967, **55**, p. 1088

8 DOBBIE, L. G. and SOMLO, P. I.: 'Precise absolute technique for measuring low-loss, symmetrical linear passive 4-terminal networks at V.H.F. and U.H.F.', *Proc. IEE*, 1963, **110**, pp. 1171–1180

9 SOMLO, P. I.: 'Cable attenuation measurement', *Proc. IRE Aust.*, 1962, **23**, p. 585

10 SLATER, J. C.: 'Microwave transmission' (McGraw-Hill, 1942), p. 37

11 GINZTON, E. L.: 'Microwave measurements' (McGraw-Hill, 1957)

12 SOMLO, P. I.: 'Localized waveguide attenuation measurement', *Proc. IREE* (*Aust*), 1969, **30**, pp. 13–15

13 GARDIOL, F. E.: 'Slotted line measurements for propagation constant in lossy waveguide', *IEEE Trans.*, 1975, **MTT-23**, pp. 317–319

14 STAUFFER, G. H., WEINERT, F. K., WEINSCHEL, B. O. and WOODRUFF, D. D.: 'Direct-reading swept-frequency slotted-line system with slope correction', *IEEE Trans.*, 1974, **IM-23**, pp. 394–399

15 OWENS, R. P.: 'Technique for the measurement of standing-wave ratios at low power level', *Proc. IEE*, 1969, **116**, pp. 933–940

16 ROBERTS, S. and VON HIPPEL, A.: 'A new method for measuring dielectric constant and loss in the range of centimeter waves', *J. Appl. Phys.*, 1946, **17**, pp. 610–616

17 BARLOW, H. M. and CULLEN, A. L.: 'Micro-wave measurements' (Constable, 1950), p. 125

18 MACKENZIE, T. E.: 'Some techniques and their limitations as related to the measurement of small reflections in precision coaxial transmission lines', *IEEE Trans.*, 1966, **IM-15**, pp. 365–374

19 HARVEY, J. A.: CSIRO National Measurement Laboratory, Australia, 1981, Private communication

20 WEINSCHEL, B. O., SORGER, G. U., RAFF, S. J. and EBERT, J. E.: 'Precision coaxial VSWR measurements by coupled sliding-load technique', *IEEE Trans.*, 1964, **IM-13**, pp. 292–300

21 CLIFTON, B. J.: 'Precision slotted-line impedance measurements using computer simulation for data correction', *IEEE Trans.*, 1970, **IM-19**, pp. 358–364

22 SORGER, G. U. and WEINSCHEL, B. O.: 'Swept frequency high resolution VSWR measuring system', Weinschel Eng., Gaithersburg, Md., Internal Rep. no. 90–117, March 1966

23 ADAM, S. F.: 'Swept-frequency SWR measurement in coaxial systems', *Hewlett-Packard J.*, 1966, pp. 14–19
24 SORGER, G. U.: 'Coaxial swept-frequency VSWR measurements using slotted lines', *IEEE Trans.*, 1968, **IM-17**, pp. 403–412
25 DUFFIN, W. J.: 'Three-probe method of impedance measurement', *Wireless Eng.*, 1952, **29**, pp. 317–320
26 SPECTOR, N.: 'Check impedance with an electronic slotted line', *Microwaves*, 1976, **15**, pp. 48–51
27 CALDECOTT, R.: 'The generalized multiprobe reflectometer and its applications to automated transmission line measurements', *IEEE Trans.*, 1973, **AP-21**, pp. 550–554
28 SAMUEL, A. L.: 'An oscillographic method of presenting impedances on the reflection-coefficient plane', *Proc. IRE*, 1947, **35**, pp. 1279–1283
29 COLE, R. S. and HONEYMAN, W. N.: 'Two automatic impedance plotters', *Elect. Eng.*, 1958, **30**, pp. 442–446
30 DE RONDE, F. C.: 'A precise and sensitive reflectometer providing automatic full-band display of reflection coefficient', *IEEE Trans.*, 1965, **MTT-13**, pp. 435–440
31 GOTTFRIED, S.: 'A millimeter wave complex reflection coefficient scanner', *IEEE Trans.*, 1973, **IM-22**, pp. 325–330
32 CHANDRA, K., PARSHAD, R. and KUMAR, R. C.: 'Measurement of impedance at microwave frequencies using directional couplers and adjustable short-circuit', *Proc. IEE*, 1967, **114**, pp. 1653–1655
33 JHA, R. K. and GARG, V. K.: 'Voltage standing wave ratio measurement by cross coupler', *Int. J. Electr.*, 1970, **29**, pp. 179–183
34 GILMER, P. E.: US Patent 2,617,853, Nov. 11, 1952
35 SOMLO, P. I.: 'Measurement of impedance at frequencies above 300 Mc', *IEEE Trans.*, 1963, **MTT-11**, pp. 87–88
36 WEINSCHEL, B. O.: Private communication, April 4, 1963

Chapter 5

Detection of reflected waves

Since an incident wave carries no information about the impedance of an unknown load, we may expect to gain some advantages over the methods of Chapter 4 by measuring only the wave reflected from the load. These advantages are particularly evident when measuring small reflection coefficients.

Circuits used to measure reflected waves may be classified as either bridges (balanced or unbalanced), or reflectometers. Bridges are designed to compare two signals – one reflected from the unknown and one from a known termination – and provide an indication of the difference between them. When the circuit is adjusted for this difference to be zero, and therefore the indication to be a null, the bridge is said to be balanced.

Reflectometers, however, provide a direct measure of the signal reflected from the unknown. The significance of this signal must be determined by comparing it with that from a known termination substituted at the measurement port.

5.1 Balanced bridges

There are several advantages in using a balanced bridge circuit to obtain a null indication from the detector. As the circuitry is adjusted and the null is approached, the sensitivity of the detector may be increased to provide higher resolution. In addition, since there is no signal incident on the detector at balance, the detector law and impedance are unimportant.

5.1.1 180° hybrid bridge

An ideal 180° hybrid (magic tee in waveguide) has the *S* parameters given in eqn. 3.36. It follows from these parameters that any signal entering port 4 will divide equally between ports 1 and 2. Signals reflected from ports 1 and 2 will divide equally between ports 3 and 4, and the signal emerging from port 3 will be proportional to the difference between the reflected signals, while that emerging from port 4 will be proportional to the sum of the reflected signals. It is therefore convenient to refer to ports 3 and 4 of a 180° hybrid as the difference

(Δ) and sum (Σ) ports, respectively, and they are denoted as such in the bridge circuit in Fig. 5.1.

In Fig. 5.1, the generator is attached to the Σ-port, and the incident signal will divide equally into the measuring arm with the unknown Γ_L attached at port 1, and the reference arm with the standard reflection coefficient attached at port 2. The difference between the reflections from these terminations will be indicated by the detector D attached to the Δ-port. When the standard reflection coefficient Γ_s has been adjusted to produce no detector output, then $\Gamma_L = \Gamma_s$ using an ideal hybrid junction.

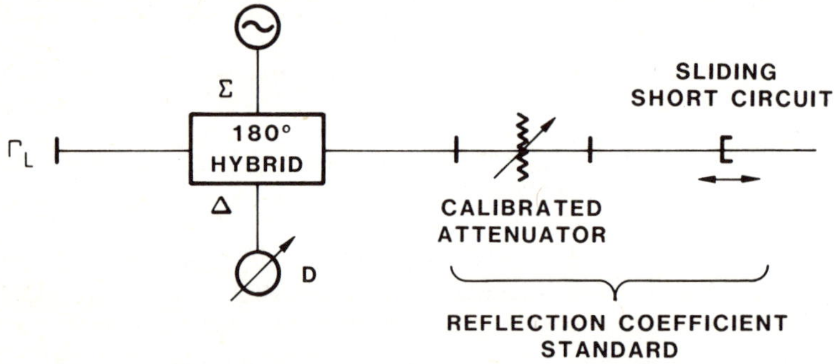

Fig. 5.1. *Simple balanced bridge circuit using a 180° hybrid junction. The detector will be nulled when the reflection coefficient standard equals* Γ_L

The flow graph representation of a general 180° hybrid driven at the Σ-port by a matched generator is shown in Fig. 5.2. In practice, the generator need not be matched, but it can be represented as such in this application since any effects of mismatch will not affect the conditions for a null at the Δ-port. The transmission from the Σ-port to the Δ-port (with a matched generator) is denoted by T_{34} where

$$\begin{aligned} T_{34} = {} & [S_{34}(L_1L_2 - \Gamma_1\Gamma_2S_{12}S_{21}) + L_1\Gamma_2S_{24}S_{32} + L_2\Gamma_1S_{14}S_{31} \\ & + \Gamma_1\Gamma_2(S_{12}S_{24}S_{31} + S_{21}S_{14}S_{32})]/[L_1L_2L_3 - L_1\Gamma_2\Gamma_3S_{23}S_{32} \\ & - L_2\Gamma_1\Gamma_3S_{13}S_{31} - L_3\Gamma_1\Gamma_2S_{12}S_{21} \\ & - \Gamma_1\Gamma_2\Gamma_3(S_{12}S_{23}S_{31} + S_{13}S_{32}S_{21})] \end{aligned} \quad (5.1)$$

and $L_i = 1 - \Gamma_iS_{ii}$. Provided the non-ideal hybrid is reciprocal ($S_{ij} = S_{ji}$) and symmetric such that $S_{34} = 0$, $S_{14} = S_{24}$, $S_{13} = -S_{23}$, $S_{11} = S_{22}$, the condition for the numerator of eqn. 5.1 to be zero and therefore there be no output at the Δ-port is $\Gamma_1 = \Gamma_2$. Therefore, a practical 180° hybrid, provided that it is reciprocal and symmetric, is as useful an impedance bridge as an ideal 180° hybrid. This is a particularly important result for a waveguide magic tee which can be made effectively symmetric by careful mechanical construction.

When the symmetry of the 180° hybrid is uncertain, or the highest measurement accuracy is sought, the hybrid may be used in a substitution mode. First the bridge is balanced with the unknown connected to port 1 and an auxiliary adjustable impedance connected to port 2. The reference standard is then substituted for the unknown and adjusted to return to the balanced condition. In this procedure, the 180° hybrid acts only as an equality indicator, and its parameters are unimportant. However, the hyrbid and the associated bridge circuitry must remain stable during the course of the substitution. An example of the use of a magic tee in the substitution mode occurs in the absolute calibration of rotary phase shifters [1].

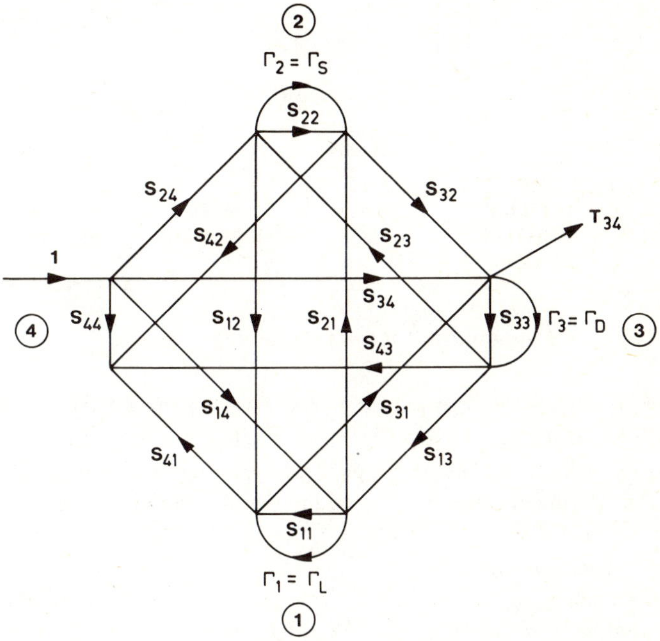

Fig. 5.2. *Flowgraph representation of a balanced bridge circuit in Fig. 5.1 driven by a matched generator*

5.1.2 90° hybrid bridge

The circuit of Parshad *et al.* [2] in Fig. 5.3 comprises two directional couplers connected in tandem, and driven by a matched generator. The main coupler divides the incident signal between the unknown and the reference termination, and sums the reflected signals. This sum is sensed by a detector connected to the auxiliary coupler, and can be made zero by adjustment of the reference termination. With a short circuit connected to the measuring port, the reference termination, comprising a calibrated attenuator and a sliding short circuit, is adjusted to produce a null response from the detector. The attenuator setting A_2 (dB) and the position of the sliding short circuit are noted. The unknown is then connected to the measuring port, and the detector nulled by changing the position of the sliding short circuit and setting the attenuator to A_1 (dB). Then $|\Gamma_L|$ is given by

$$|\Gamma_L| = 10^{(A_1 - A_2)/10} \tag{5.2}$$

and from the change in the position of the sliding short circuit and a knowledge of the change in phase shift through the attenuator between A_1 and A_2, the phase of the unknown reflection can be determined.

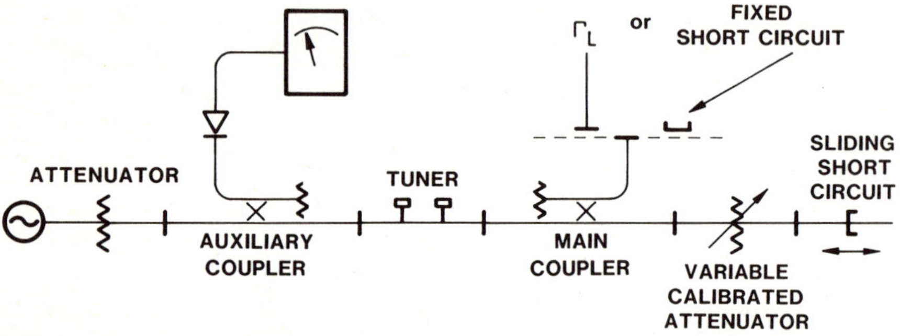

Fig. 5.3. *The 90° hybrid bridge circuit of Parshad et al.* [2]

The tuner in Fig. 5.3 is introduced to improve the directivity of the auxiliary coupler, and is adjusted to produce a null at the detector when both the unknown and reference ports of the main coupler are terminated in matched loads. This type of directivity improvement is discussed in detail in Section 5.3.2.

5.1.3 G-R admittance bridge

A coaxial admittance bridge developed by the General Radio (G-R) Company [3, 4] and usable from 20 MHz to 1·5 GHz is shown schematically in Fig. 5.4. The currents flowing in the three coaxial branch lines are driven from the common junction, and are sampled by three independently rotatable electrostatically shielded loops whose outputs are connected in parallel. Two of the branch lines are terminated in conductance and susceptance standards, and the third in the unknown admittance. The conductance standard G_s is chosen to be equal to the (real) characteristic admittance $Y_0 = 0{\cdot}02$ S, and the susceptance standard B_s is chosen to ensure that the admittance under the loop in the susceptance branch is $\pm jY_0$. The susceptance standard is realized at the higher frequencies as a coaxial line of effective length $\lambda/8$ terminated in an open or short circuit, and at the lower frequencies as an air-filled capacitor.

Because the voltage induced in a loop is proportional to the product of the current flowing in the coaxial line beneath the loop and the cosine of the rotation angle, it is possible to rotate the loops in the two standard branches until the detector indicates a null. The real and imaginary parts of the unknown admittance may then be read directly from the rotation scales attached to the standard conductance and susceptance branches, respectively. The rotation of the loop in the branch of the unknown affects a common scaling factor.

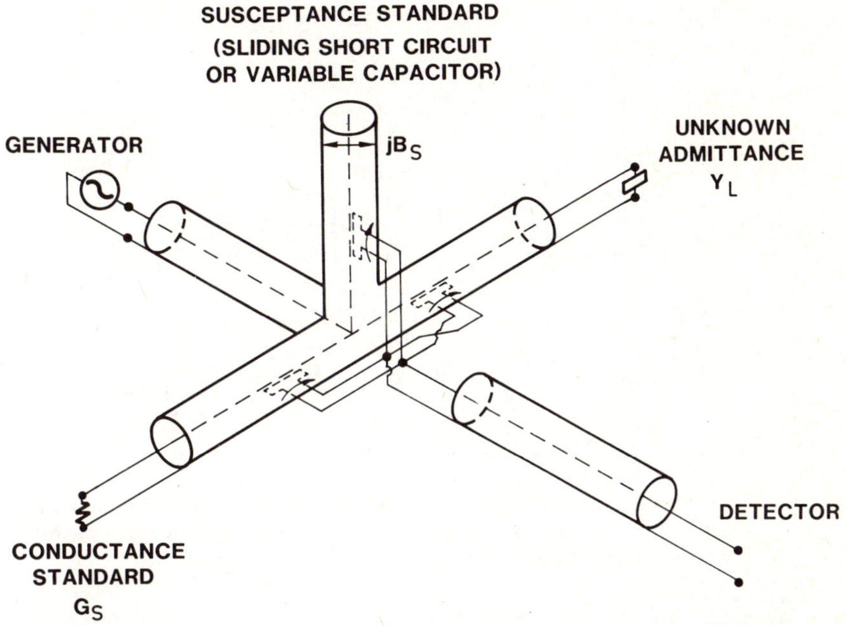

Fig. 5.4. *General Radio Company admittance bridge. The magnetic pick-up loops are rotated in the standard conductance and susceptance arms until a null is produced at the detector* [3]

The admittance measured by this bridge is that under the loop in the branch of the unknown, and a transmission line transformation is required to calculate the admittance in the plane of the unknown. The uncertainty expected in using the G-R bridge is about 3% rising to 5% at the upper frequency limit. The operation of the bridge is limited at the lower frequencies by the difficulty in providing the susceptance standard, and at the higher frequencies by the finite distances between the loops and the common junction.

For more precise measurements, the bridge may be used as an equality detector by employing a substitution technique similar to that described in Section 5.1.1 for 180° hybrid bridges. An example of such use is the precision impedance measurements which can be made by first connecting a tuner terminated in a nominally matched load to the measuring port of the bridge, and adjusting the tuner until the combination is reflectionless (Fig. 5.5). This condition is attained when a $\lambda/4$ length of precision coaxial line can be inserted between the tuner and the bridge without unbalancing the bridge. The matched combination is then removed from the measuring port, and replaced by a calibratable tuner terminated in the unknown admittance. This tuner is adjusted until the unknown admittance is transformed to a match at the measuring port, as evidenced by restoration of the bridge balance. The admittance of the unknown may then be deduced from the calibration of the tuner [5].

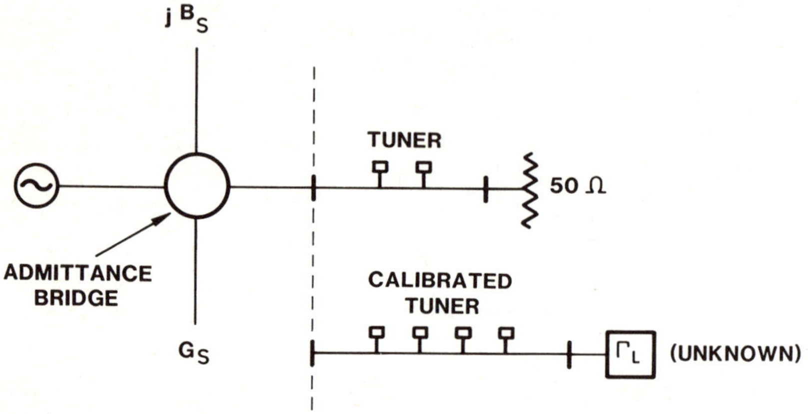

Fig. 5.5. *Substitution impedance measurement using a G-R admittance bridge as an equality detector. The unknown impedance is determined from the settings of the calibrated tuner required to provide a match at the plane of substitution*

5.1.4 Woods bridge

There are available many designs of bridges made from discrete components, but because of the residual impedances associated with the components they are limited to an upper frequency of around 300 MHz, and cannot be classified as microwave instruments. However, because of the consideration given to high frequency performance and high precision, the bridge designed by Woods [6] needs special mention.

The basic circuit of the Woods bridge given in Fig. 5.6 employs a twin tee circuit which allows the source, detector, and unknown to have a common earth point. By careful design, all the residual effects of the series elements of the bridge are eliminated except the variation of G_s with frequency, which, because of the dual nature of the bridge, can be determined using terminals T_{xa} and T_{xb}.

The frequency range of the bridge is from 3 MHz to 300 MHz with a claimed accuracy of around 0·2% at 200 MHz, which classifies it as a precision measuring instrument.

5.2 Unbalanced bridges

When it is sufficient to measure only the magnitude of the reflection coefficient (or VSWR), considerable equipment simplification can be achieved by using unbalanced bridges.

5.2.1 Resistive bridge

A modified Wheatstone bridge is shown in Fig. 5.7*a* with the branch resistors equal to the chosen characteristic impedance (usually 50 ohms). If the detector is of high impedance, as in the conventional useage of a Wheatstone bridge at low frequencies, the (normalized) impedance Z'_m seen looking into the measuring port is not a match unless the generator impedance Z'_G is zero, which is impractical at microwave frequencies.

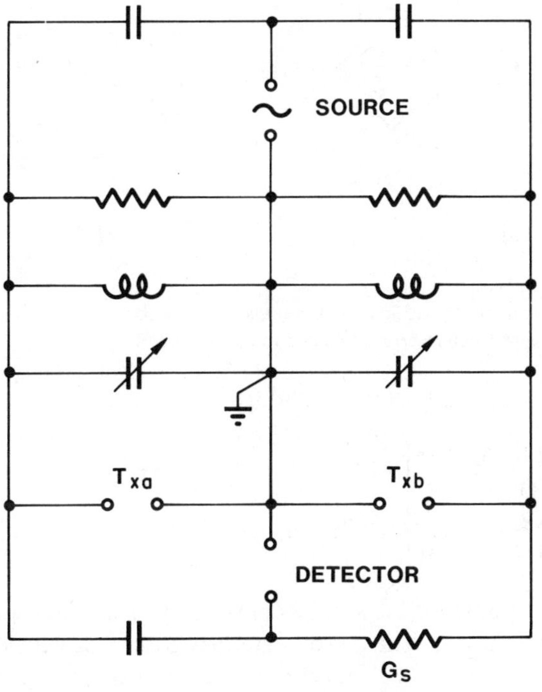

Fig. 5.6. *Circuit of the Woods' bridge* [6]

However, if the detector is matched ($Z'_D = 1$), as indicated in Fig. 5.7*b*, then

$$Z'_m = \frac{3 + 5Z'_G}{5 + 3Z'_G} \tag{5.3}$$

and provided the generator is matched ($Z'_G = 1$), a match is presented at the measurement port. Such a bridge may be constructed in compact form using integrated resistors to avoid the problems of finite lead lengths. When an unknown impedance Z'_L is connected to the measuring port, the ratio V, of voltage across the detector to bridge potential E, is related to Γ_L by

$$|\Gamma_L| = 8|V| \tag{5.4}$$

illustrating that, just as for a simple reflectometer using an ideal directional coupler, the detector voltage is proportional to the magnitude of the reflection coefficient of the unknown. Often the detector is built into the bridge, and if it is operated in the square-law region the detector output will be proportional to $|\Gamma_L|^2$.

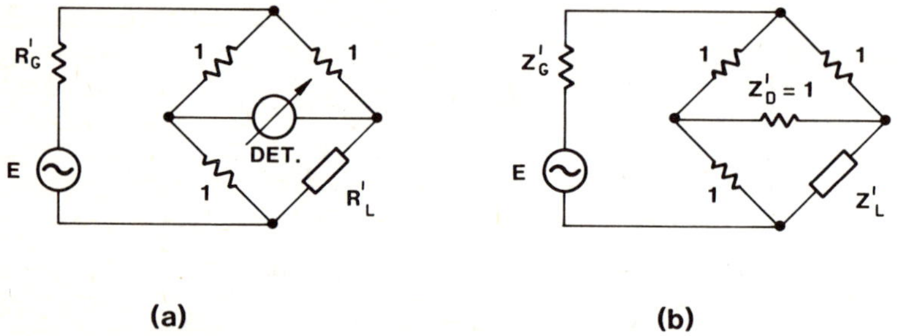

Fig. 5.7. *Resistive bridges*
a Wheatstone bridge using a high impedance detector
b Microwave resistive bridge where a matched detector is used

The S parameter matrix for the ideal bridge is

$$S = \tfrac{1}{2}\begin{bmatrix} 0 & 1 & 0 \\ 1 & 0 & 1 \\ 0 & 1 & 0 \end{bmatrix} \tag{5.5}$$

where the generator port is designated as port 1, the measuring port as port 2, and the detector port as port 3. The matrix describes a matched but lossy 3-port with ports 1 and 3 decoupled, but with equal transmissions from port 2 to ports 1 and 3.

If the generator and detector are mismatched with reflection coefficients Γ_G and Γ_D, respectively, the reflection coefficient seen looking into the measuring port is $\Gamma_m = (\Gamma_G + \Gamma_D)/4$, and

$$V = \frac{\Gamma_L}{8} \frac{(1+\Gamma_D)(1-\Gamma_G)}{1-\Gamma_L\Gamma_m} \tag{5.6}$$

In order to find the effect of Γ_m on the value measured for $|\Gamma_L|$, we first let $|V_{sc}|$ denote the mean of the maximum and minimum magnitudes of the detector signal as the position of a calibrating sliding short circuit connected to the measuring port is adjusted. Then when Γ_L is connected, the detector signal is

$$\frac{|V_L|}{|V_{sc}|} = |\Gamma_L| \frac{1 - |\Gamma_m|^2}{|1 - \Gamma_L \Gamma_m|} \tag{5.7}$$

From this it can be shown that the bridge sensitivity to generator and detector mismatches is similar to that of a simple reflectometer (Section 5.3.1) having a coupling of 6 dB, and with a 6 dB attenuator between the generator and port 1.

Apart from its inherent simplicity, the main advantage of the resistive bridge is its potential for wideband operation. Commercially available bridges operate to 20 GHz with an effective directivity of 36 dB and are suitable for wideband swept-frequency operation. For swept-frequency operation, it is often convenient to display the detector output as the vertical deflection of a CRO, with the horizontal deflection being a linear function of frequency. The variation of $|\Gamma_L|$ with frequency may then be found directly by comparing the trace with that of a reference reflection. Special techniques to separate the wanted from the unwanted (internal) reflections may be applied [7].

Disadvantages of a bridge equipped with a square-law detector are the low sensitivity when a small reflection coefficient is being measured, and that only the magnitude of the reflection coefficient is found.

5.2.2 Stripline VSWR bridge

Furukawa and Aoyagi [8] devised the VSWR bridge consisting of a TEM stripline containing a tee junction, as shown oriented with its axis in the y-direction in Fig. 5.8. The arms of the tee lie in the x-direction and one is terminated in a normalized susceptance standard B_s' with the other terminated in the unknown Y_L'. An electrostatically shielded, rotatable loop is suspended above the junction. When the lengths of the arms are short compared to the wavelength, the currents flowing in them are approximately proportional to the terminating admittances, and therefore the total currents in the x- and y-directions at the junction are

$$I_x \propto Y_L' - jB_s', \quad I_y \propto Y_L' + jB_s' \tag{5.8}$$

If the plane of the loop makes an angle ψ with the x-direction, the voltage V induced in the loop is proportional to

$$V \propto I_x \cos(\psi) + I_y \sin(\psi) = (Y_L - jB_s)\cos(\psi) + (Y_L + jB_s)\sin(\psi) \tag{5.9}$$

and the ratio of the maximum induced voltage to the minimum induced voltage as the loop is rotated is the VSWR given in eqn. 1.25, where Γ_L is the reflection coefficient which Y_L possesses in a line of characteristic admittance B_s. The argument, ϕ_L, of Γ_L is given by

$$\phi_L = 2[\psi_{min} - \pi/4] \tag{5.10}$$

where ψ_{min} is the value of ψ at a detected minimum. It is not necessary to know the value of B_s in order to find ϕ_L.

The bridge is claimed to be useful up to 2 GHz, with the overall uncertainty of less than 3% mainly caused by the finite distance from the junction to the terminations. It may be used for the measurement of active admittances (those with a negative real part), and suitable resistive loading of the standard susceptance ensures absolute stability.

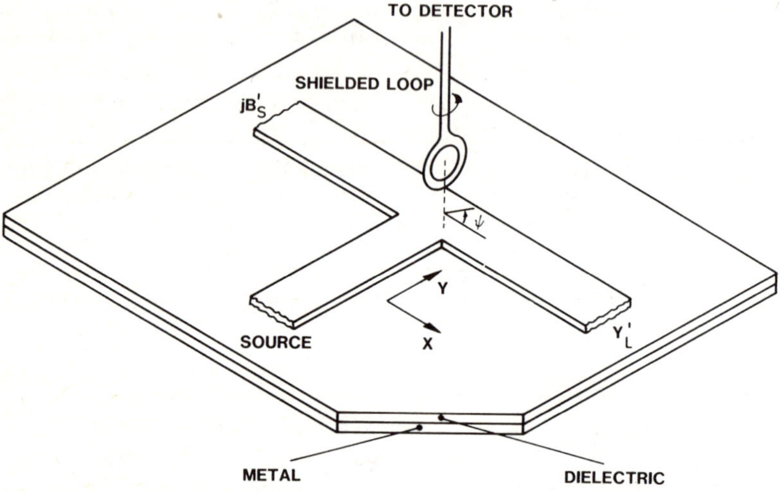

Fig. 5.8. *Stripline impedance measuring system of Furukawa and Aoyagi* [8]
The VSWR is determined by the ratio of the maximum to minimum loop pick-up as the loop is rotated, and the phase can be determined from the rotation angle of the loop at the minimum

5.3 Reflectometry using directional couplers

The ability of a directional coupler to selectively couple the forward or reflected waves in a transmission line render it one of the most useful components available for impedance measurements. In a practical coupler, the ability to discriminate between forward and reflected waves is not perfect (directivity is finite), and is affected by the coupler terminations. Various circuits using directional couplers as reflectometers have been designed to increase the effective directivity of the coupler and enhance the ability to separate the reflected from the incident signal.

5.3.1 Simple reflectometer

The simple reflectometer in Fig. 5.9 is probably the most commonly used microwave impedance measuring circuit. With port 3 terminated in a square-law detector, and port 4 in a nominally matched load, the detector output is approximately propotional to $|\Gamma_L|^2$, and the constant of proportionality may be found by connecting known reflection standards to the measuring port. The detected signal may be displayed directly on a d.c. meter, or the generator may be amplitude (most often square-wave) modulated and a.c. amplification used to increase the sensitivity without d.c. offset errors.

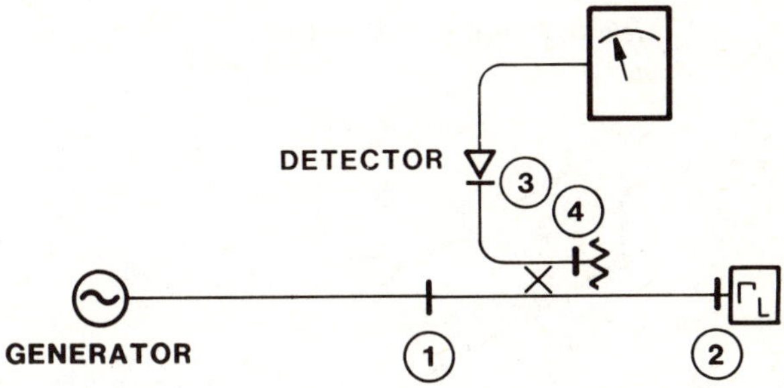

Fig. 5.9. *Circuit of the simple reflectometer using a directional coupler*

The primary limitation of the simple reflectometer is the finite directivity (leakage) of the directional coupler. Consider the circuit in Fig. 5.9 comprising a matched generator, matched detector, and a high quality directional coupler of directivity coefficient d, with port 4 terminated in a matched load. When a termination with reflection coefficient Γ_L is connected to port 2, if follows from the S parameter description of the coupler that the signal incident on the detector is (Fig. 3.25b)

$$b_3 = S_{31} + \frac{S_{32}S_{21}\Gamma_L}{1-S_{22}\Gamma_L} = S_{32}\left[\frac{S_{31}}{S_{32}} + \frac{S_{21}\Gamma_L}{1-S_{22}\Gamma_L}\right] \tag{5.11}$$

With S_{22} small, this signal may be approximated by the sum of two signals with magnitudes proportional to d^{-1} and $t|\Gamma_L|$, respectively, where $t = \sqrt{(1-c^2)}$ (Section 3.3.4.2). These two signals are summed vectorially at the detector, and the d^{-1} component causes a deviation from linearity of detector input with $|\Gamma_L|$. In particular, when $|\Gamma_L|$ is comparable to d^{-1} the reflectometer is subject to errors of up to ± 100%. For example, a simple reflectometer with 40 dB directivity ($d = 10^2$) cannot be used to reliably measure reflection coefficient magnitudes of the order of 0·01 or less.

If the generator and detector connected to a perfect coupler are mismatched with reflection coefficients Γ_G and Γ_D, respectively, the reflection coefficient seen looking into the measuring port, and the detector input, are

$$\begin{aligned} \Gamma_m &= t^2\Gamma_G - c^2\Gamma_D, \\ b_3 &= \frac{-jct\Gamma_L}{1-\Gamma_L\Gamma_m} \end{aligned} \tag{5.12}$$

Let $|b_{sc}|$ denote the mean of the maximum and minimum magnitudes of b_3 as the position of a calibrating sliding short circuit connected to the measuring port is adjusted. Then when Γ_L is connected

$$\frac{|b_L|}{|b_{sc}|} = |\Gamma_L| \frac{1 - |\Gamma_m|^2}{|1 - \Gamma_L \Gamma_m|} \tag{5.13}$$

which should be compared with eqn. 5.7 for a bridge.

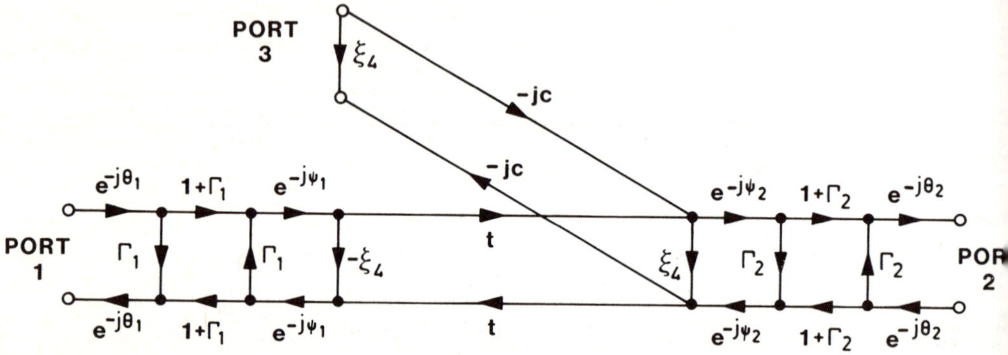

Fig. 5.10. *Flow graph representation of a reciprocal, lossless, directional coupler with the fourth port internally terminated*

If the practical lossless coupler represented by the flow graph in Fig. 3.24*b* is terminated on port 4 with a reflection coefficient of Γ_T, and this port is made inaccessible, the flow graph of the resulting three-port may be simplified to that in Fig. 5.10 from which

$$d^{-1} = t \left| \frac{(1 + \Gamma_1)(\xi_4 + \Gamma_2')}{(1 + \Gamma_2)(1 + \xi_4 \Gamma_1')} \right| \tag{5.14}$$

where $\Gamma_i' = \Gamma_i e^{-j2\psi_i}$, and

$$\xi_4 = \frac{\Gamma_4 + 2\Gamma_4 \Gamma_T' + \Gamma_T'}{1 - \Gamma_4 \Gamma_T'} e^{-j2\psi_4}; \quad \Gamma_T' = \Gamma_T e^{-j2\theta_4} \tag{5.15}$$

When the three-port is driven by a generator with reflection coefficient Γ_G at port 1, and a detector with reflection coefficient Γ_D is attached to port 3, the signal T_{31} incident on the detector is

$$T_{31} = \frac{-jct(1 + \Gamma_1)(\xi_2 + \xi_4)\, e^{-j(\theta_1 + \psi_1)}}{(1 - \Gamma_1 \Gamma_G')[(1 - \xi_1 \xi_2)(1 - \Gamma_D \xi_4) + c^2(\xi_2 + \xi_4)(\xi_1 + \Gamma_D)]} \tag{5.16}$$

where ξ_1, Γ_G' and ξ_2, Γ_L' are given by expressions similar to those in eqn. 5.15 but appropriate to ports 1 and 2, respectively. The quantity ξ_i may be interpreted as the reflection coefficient seen from the coupler core when looking towards the terminated ith port. The effective directivity of the reflectometer may be determined after associating Γ_G and Γ_D with ports 1 and 3 of the coupler

$$d^{-1} = t\left|\frac{(1+\Gamma_1)(\xi_4+\Gamma_2')}{(1+\Gamma_2)(1+\xi_1\xi_4)(1-\Gamma_1\Gamma_G')}\right| \tag{5.17}$$

which is independent of Γ_D.

A qualitative assessment of the effect of coupler port mismatches on the performance of a simple reflectometer may be made after approximating a high quality practical coupler by an ideal coupler. Then from eqns. 5.16 and 5.17

$$T_{31} = \frac{-jct(\Gamma_L+\Gamma_T)}{(1-\Gamma_L\Gamma_G)(1-\Gamma_D\Gamma_T)+c^2(\Gamma_L+\Gamma_T)(\Gamma_G+\Gamma_D)} \tag{5.18}$$

and

$$d^{-1} = t\left|\frac{\Gamma_T}{1+\Gamma_G\Gamma_T}\right| \tag{5.19}$$

It is apparent that if $|\Gamma_T|$ is small, the directivity of the reflectometer is large, and the measurement of small values of $|\Gamma_L|$ is possible in the presence of generator and detector reflections. However, even if $\Gamma_T = \Gamma_D = 0$ giving

$$T_{31} = \frac{-jct\Gamma_L}{1-t^2\Gamma_L\Gamma_G} \tag{5.20}$$

the generator mismatch may cause a large error when measuring large values of $|\Gamma_L|$. In particular, if port 2 is terminated in a sliding short circuit ($\Gamma_L = e^{j\phi_L}$) the ratio of maximum to minimum signals at the detector as ϕ_L is varied is

$$\left|\frac{T_{31\,(\mathrm{max})}}{T_{31\,(\mathrm{min})}}\right| = \frac{1+t^2|\Gamma_G|}{1-t^2|\Gamma_G|} \tag{5.21}$$

which provides a method for measuring generator impedance $|\Gamma_G|$ reminiscent of that in Section 4.2.4.1.

5.3.2 Tuned reflectometer

The tuned reflectometer of Engen and Beatty [9] is a circuit which converts a simple reflectometer into an ideal reflectometer at one frequency.

If two lossless two-ports are connected in tandem, the combination is also a lossless two-port. Consequently, if lossless tuners are connected to ports 1 and 2 of a practical lossless coupler, the coupler–tuner combination can also be represented by the flow graph in Fig. 5.10 and the transmission T_{31} in eqn. 5.16 describes the signal incident on the detector connected to port 3 when the generator and unknown are connected to the new ports 1 and 2, respectively, as shown in Fig. 5.11. Unlike the case of the simple reflectometer, it is now possible to vary the two-port parameters by adjusting the tuners.

Examination of eqn. 5.16 reveals that T_{31} is in the form of a bilinear transformation

$$T_{31} = \frac{A + B\Gamma_L}{C + D\Gamma_L} \tag{5.22}$$

where the complex quantities A, B, C, D are independent of Γ_L. The purpose of the tuners is to ensure that $A = D = 0$, and therefore that $T_{31} \propto \Gamma_L$.

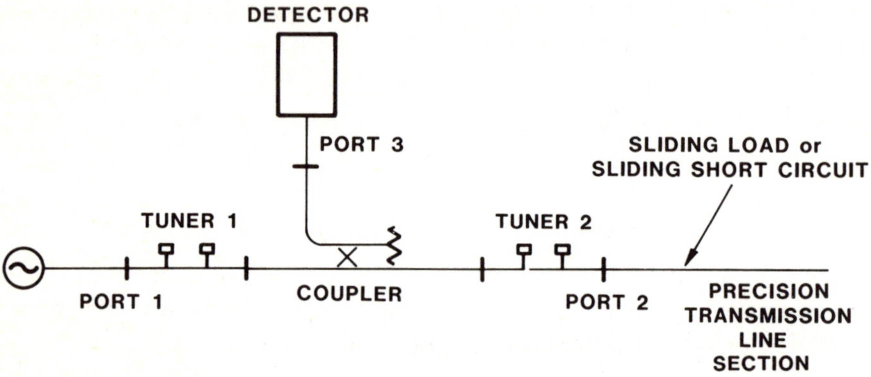

Fig. 5.11. *Tuned reflectometer of Engen and Beatty [9]. By adjusting the two tuners, a practical imperfect coupler and source are transformed into an ideal reflectometer at the operating frequency*

The condition $A = 0$ is characterized by the absence of detector output ($T_{31} = 0$) when a reflectionless termination ($\Gamma_L = 0$) is connected to port 2. In terms of the parameters in eqn. 5.16, the condition is met if tuner 2 is adjusted until

$$\xi_2 \Big|_{\Gamma_L = 0} = -\xi_4, \quad \text{i.e.} \quad -\Gamma_2 e^{-j2\psi_2} = \xi_4 \tag{5.23}$$

It then follows from eqn. 5.16 that $D = 0$ provided that tuner 1 can be adjusted until

$$-\xi_1 e^{-j2\psi_2} = \frac{\Gamma_2(1 + \Gamma_2 z) - c^2(1 + \Gamma_2)^2 z}{(2\Gamma_2 + 1)(1 + \Gamma_2 z) - c^2(1 + \Gamma_2)^2} \tag{5.24}$$

where $z = \Gamma_D e^{-j2\psi_2}$. Since the magnitude of the right-hand side of eqn. 5.24 can be shown to be less than unity, this adjustment is possible.

In practice, because a reflectionless termination is not available, a termination with low reflection is slid inside a length of precision transmission line as shown in Fig. 5.11, and tuner 2 adjusted to minimize the resulting level variations (ripple) of the detector output. The sliding load is then replaced by a sliding short circuit, and tuner 1 adjusted to obtain minimum ripple on the detector output. Because the sliding load is not reflectionless, the tuner adjustments are not independent and the tuning procedure should be repeated until no ripple is apparent as the terminations are moved [9]. It is desirable that the precision transmission line of Fig. 5.11 be left attached to the tuned reflectometer for subsequent measurements, as the tuning procedure has also compensated for the reflection caused by the connector joining it to tuner 2.

Because of attenuation in the precision transmission line, the detector output will not be constant as the short circuit is moved, but can be made so by using electronic normalizing circuits [10].

In order to realize the potential accuracy of the tuned reflectometer, the sliding load and short circuit should be of high quality such that the magnitudes of their reflection coefficients do not change as they are moved in the precision line. The signal source should be frequency stabilized, preferably by phase-locking, and should be levelled. A sensitive detector with high resolution is needed if the ripples are to be successfully tuned out.

Use of a microwave attenuation measuring facility [11] in place of the detector offers high resolution during the tuning procedure, and accurate signal measurement when the reflectometer is in use. Such an arrangement was used by Somlo [12] who moved a short circuit across the junction between two waveguides of different material, and observed directly the waveguide loss and flange loss.

The major disadvantage of the tuned reflectometer is that retuning is necessary for each measurement frequency.

5.3.3 Magnified response reflectometer

At the same time that Engen and Beatty [9] described the tuned reflectometer in which the leakage signal associated with finite directivity is cancelled by external tuners, Beatty [13] described a reflectometer designed to utilize the leakage signal.

For a practical high-quality coupler, $|S_{22}| \ll 1$, and the expression in eqn. 5.11 for the signal incident on the detector may be written as

$$b_3 \simeq A(\Gamma_l + \Gamma_L) \tag{5.25}$$

where $A = S_{32}S_{21}$ is a complex constant, and the leakage signal Γ_l which is always incident on the detector has a magnitude of

$$|\Gamma_l| \simeq \frac{1}{td} \tag{5.26}$$

If the magnitude of Γ_L is similar to $|\Gamma_l|$ and the phase of Γ_L can be varied by sliding the unknown along a transmission line connected to the measuring port (or introducing a perfect phase shifter), as shown in Fig. 5.12, the amplified output of the detector can range from near zero to large values. In this way, this reflectometer provides a magnified response (in decibels) to small reflected signals which could not be distinguished using the simple reflectometer.

If both the magnitude and phase of Γ_L can be adjusted until the detector output is zero, then $|\Gamma_l| = |\Gamma_L|$ and the coupler directivity can be determined from eqn. 5.26, provided Γ_L is known.

If no auxiliary method is available to determine the adjusted value of Γ_L, the termination should be moved $(\lambda_g/4)$ along the transmission line to the position of maximum detector output given by

$$|b_3^{\max}| = 2|A||\Gamma_L| \tag{5.27}$$

A short circuit is then attached to the measuring port and, provided the generator port is well-matched and the leakage is small, the resulting detector output b_3^{sc} is used to find the reflectometer directivity from

$$d \simeq \frac{2|b_3^{sc}|}{t|b_3^{\max}|} \tag{5.28}$$

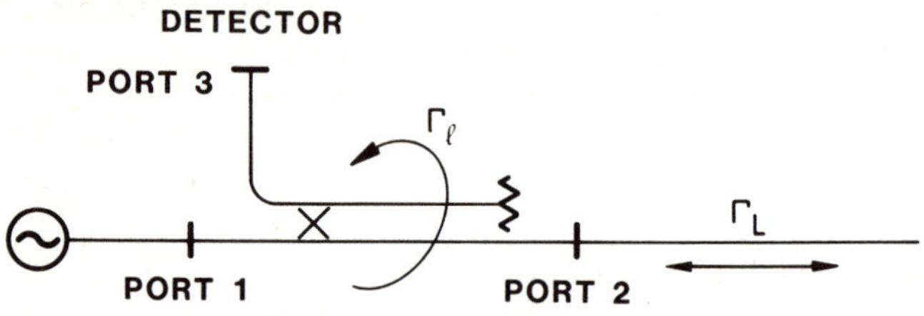

Fig. 5.12. *Magnified response reflectometer of Beatty [13]. When the unknown can be slid along a uniform transmission line, the interference of its reflection with the leakage signal is detected*

5.4 Frequency-swept reflectometry

It is tedious to make separate measurements at many different frequencies in order to characterize the unknown across a broad frequency range. This is particularly true if the unknown is to be adjusted to some predetermined specification, necessitating repeated measurements to ascertain the effect of an adjustment. With the availability of swept-frequency generators, the possibility of immediate broadband measurements, and an immediate indication of the effect of an adjustment, has much appeal.

Reference has already been made in Section 5.2.1 to the use of resistive bridges as frequency-swept reflectometers. Many of the directional coupler reflectometers discussed in the following sections may instead be based on resistive bridges.

5.4.1 Simple frequency-swept reflectometer

If the simple reflectometer in Fig. 5.9 incorporates a broadband directional coupler, then, subject to the limitations discussed in Section 5.3.1, it may be used for swept-frequency measurements. The detector output can be displayed as the vertical deflection on an oscilloscope, with the horizontal deflection made proportional to frequency.

The calibration of the display is achieved by recording the traces obtained when known terminations are connected to the measurement port. A set of such traces is shown in Fig. 5.13, and illustrates the poor resolution for small reflection coefficients when a square-law detector is used. The oscillations on the $|\Gamma_c| = 1$ calibration curve are evidence of interference between the calibrating short circuit

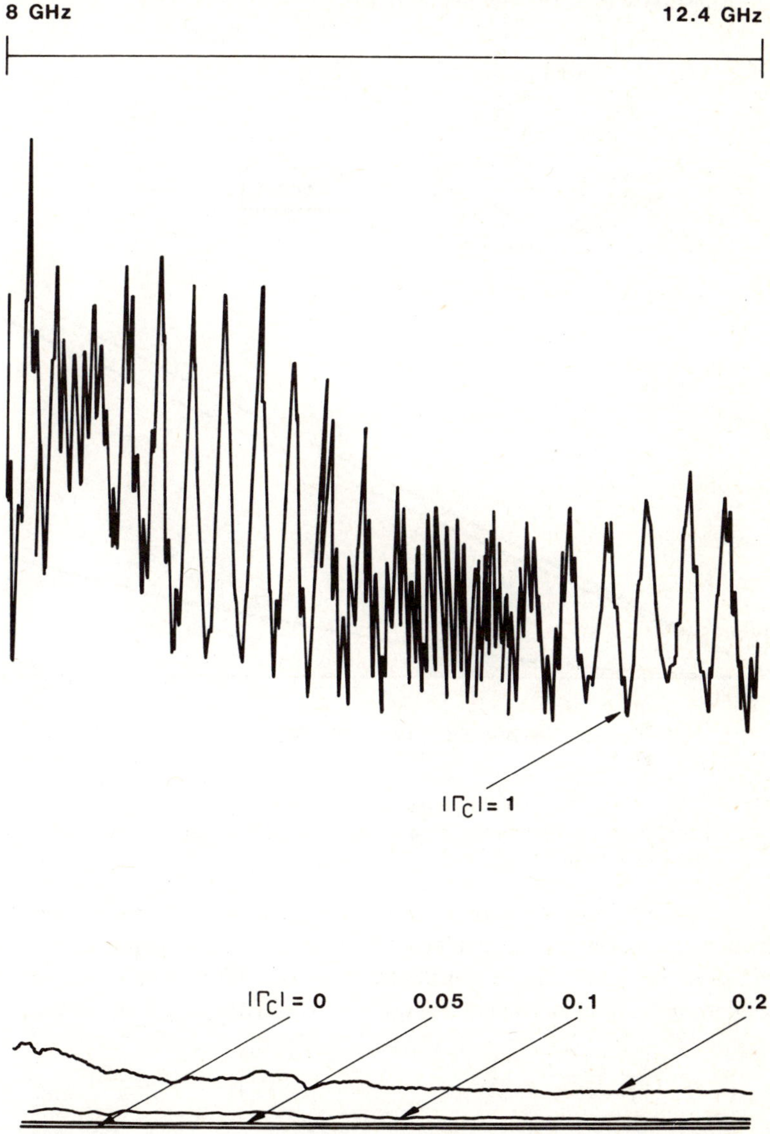

Fig. 5.13. *Results obtained with the frequency-swept simple reflectometer of Fig. 5.9.* Note that the resolution is poor for small reflections because a square-law detector was used

and the mismatched generator, and could be reduced by inserting a quality isolator between generator and coupler. The frequency of the oscillations depends on the electrical distance between the generator and coupler. If the frequency response of the coupler is constant, the introduction of signal levelling would maintain more constant sensitivity over the frequency range.

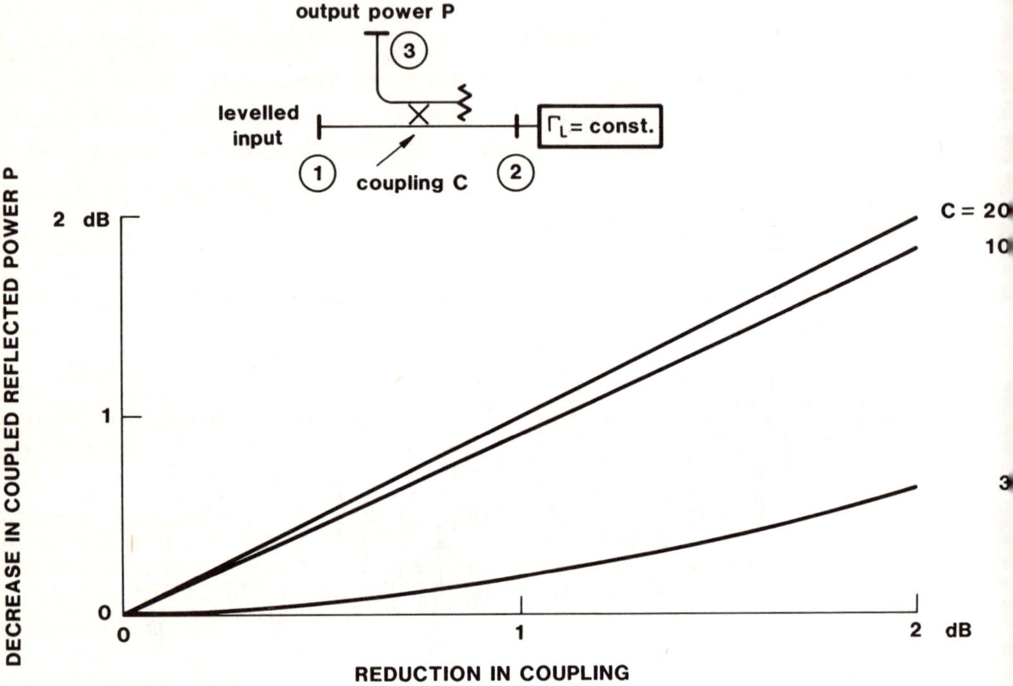

Fig. 5.14. *Effect of coupling variations in a simple reflectometer*
Note that a simple reflectometer using a 3 dB coupler is the most tolerant to coupling variations

When the frequency response of an ideal coupler is not constant, the variation in the power incident on a matched detector connected to port 3 is shown in Fig. 5.14 for the conditions of constant (levelled) input at port 1, and a frequency independent reflection coefficient at port 2. This variation is given by

$$R = \frac{\alpha(1 - \alpha P')}{1 - P'} \tag{5.29}$$

where P' and α are as defined in Section 3.3.4.3. When such a frequency-swept reflectometer is used to measure the reflection coefficient of an unknown termination, the output of the matched detector must be corrected to account for this variation. The overall frequency response of the reflectometer, including the coupler frequency variation, is readily determined after connecting a frequency independent reflecting termination, such as a short circuit, to the measuring port.

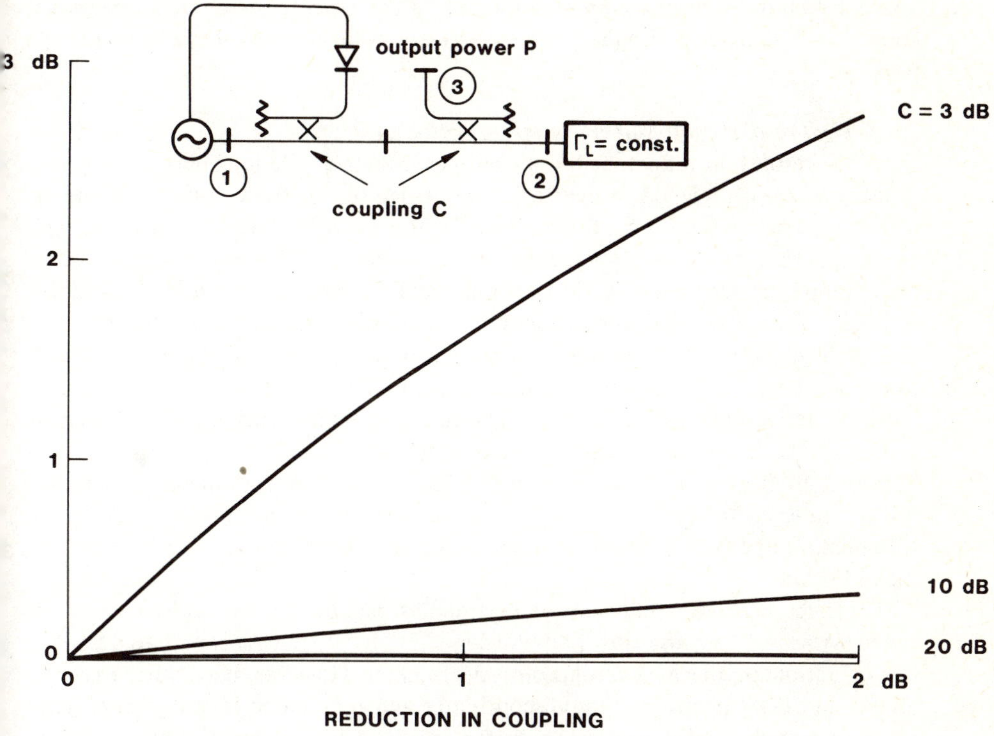

Fig. 5.15. *Effect of coupling variations in a dual-coupler reflectometer*
Note that a dual-coupler reflectometer using 3 dB couplers is the least tolerant to coupling variations

5.4.2 Dual coupler frequency-swept reflectometer

If two ideal couplers with identical frequency characteristics are connected as shown in Fig. 5.15, and the first used in a levelling circuit, the effect of coupler frequency response on the power delivered to a matched detector connected to port 3 is given by

$$R = \frac{(1-\alpha P')^2}{(1-P')^2} \tag{5.30}$$

where P' and α are as defined in Section 3.3.4.3. This variations, shown in Fig. 5.15 for a frequency independent termination connected to port 2, is seen to be much reduced in comparison with the simple reflectometer variation in Fig. 5.14, provided that the coupling coefficient is small.

The dual directional coupler connection allows the reflection coefficient of an unknown termination to be displayed directly as the y-deflection of an oscilloscope or X-Y recorder, with the x-deflection being proportional to frequency, without the need to apply a frequency correction to the detector output. The display can be scaled by connecting a known termination to the reflectometer. Dual directional couplers with matched frequency responses are available commercially in a single package.

5.4.3 *High resolution frequency-swept reflectometer*

The high resolution reflectometer, sometimes known as the long-line or delay-line reflectometer, is designed to overcome the difficulties of finite coupler directivity which become apparent with the simple reflectometer when $|\Gamma_L|$ is small. The high resolution reflectometer is characterized by the electrically long length of transmission line between the directional coupler and the measuring port [14], as shown in Fig. 5.16*a*. The long line is reminiscent of the apparatus of de Ronde [15] discussed in Section 4.2.3.1, and its purpose is to provide a variable phase shift with frequency between the signal Γ_L reflected from the termination and that Γ_r from a reference reflection near the coupler. The resulting interference signal is displayed as the y-axis deflection of an oscilloscope, with the x-axis deflection being proportional to frequency. Unlike the magnified response reflectometer, which also utilizes interfering signals, it is not necessary to move the unknown along the transmission line or to introduce a (perfect) phase shifter between the coupler and the unknown.

The reference reflection may be regarded as part of the lossless two-port associated with port 2 in the flow graph of Fig. 5.10, in which case eqn. 5.16 describes the operation of the high resolution reflectometer. However, it is rather easier to understand this operation if the discontinuity and terminated long line of length l are represented by the flow graph in Fig. 5.16*b*, which reduces to an equivalent reflection coefficient of Γ_e at port 2 of the coupler, where

$$\Gamma_e = \frac{\Gamma_r + \Gamma_L e^{-j2\theta}(1 + 2\Gamma_r)}{1 - \Gamma_r \Gamma_L e^{-j2\theta}}, \qquad \theta = \frac{2\pi l}{\lambda_g} \tag{5.31}$$

When $|\Gamma_r|$ and $|\Gamma_L|$ are small, we may approximate Γ_e by

$$\Gamma_e \simeq \Gamma_r + \Gamma_L e^{-4\pi l/\lambda_g} \tag{5.32}$$

from which it can be seen that sweeping the frequency such that the (guide) wavelength decreases from λ_1 to λ_2 will cause $|\Gamma_e|$ to have n maxima, where

$$n \simeq 2l\left(\frac{1}{\lambda_2} - \frac{1}{\lambda_1}\right) \tag{5.33}$$

Further, if we temporarily make the assumption that the coupler leakage signal may be combined with Γ_r, then as the frequency is swept the difference between the maximum and minimum outputs of a square-law detector attached to port 3 is

$$V_{\text{diff}} \propto (|\Gamma_r| + |\Gamma_L|)^2 - (|\Gamma_r| - |\Gamma_L|)^2 = 4|\Gamma_r||\Gamma_L| \tag{5.34}$$

Therefore, to a first order of accuracy, this difference is proportional to $|\Gamma_L|$ and the sensitivity is proportional to $|\Gamma_r|$. When $|\Gamma_L|$ is near zero, a choice of $0{\cdot}05 \leqslant |\Gamma_r| \leqslant 0{\cdot}2$ is appropriate.

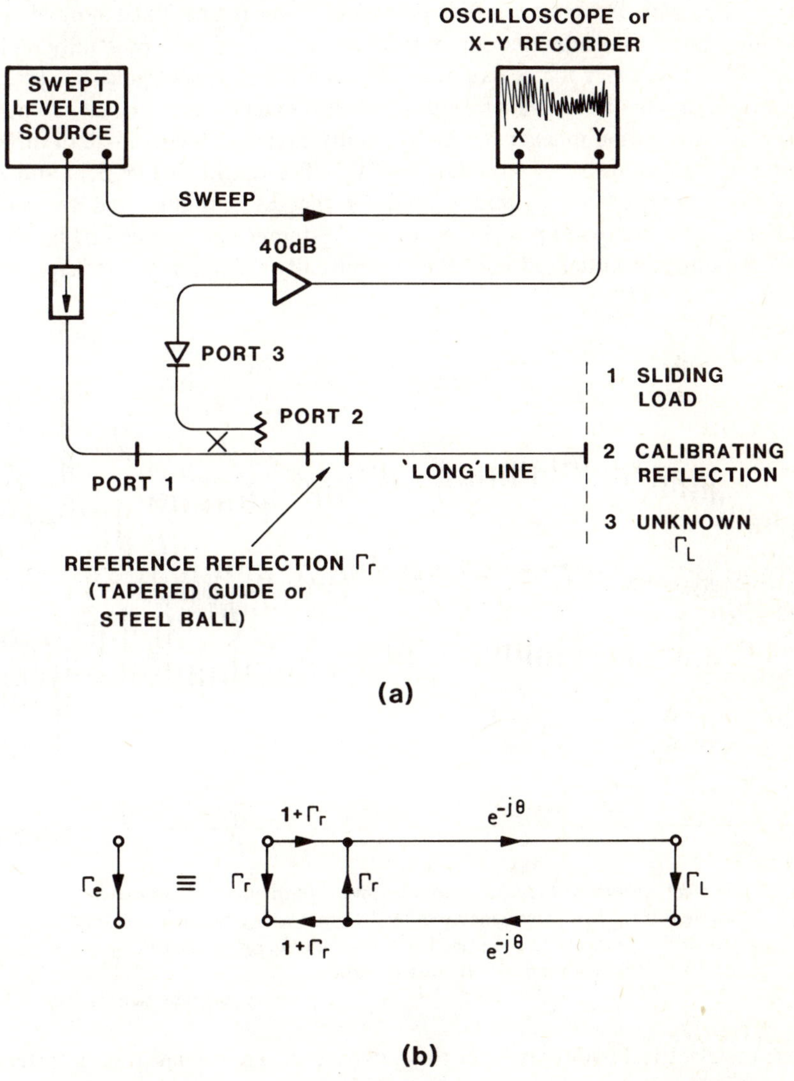

Fig. 5.16. *High resolution swept-frequency reflectometer of Hollway and Somlo* [16]
a Circuit. As the frequency is swept, a high frequency interference signal is produced because of the large separation between the unknown and reference reflections
b Signal flow graph representation of the circuit connected to port 2

The high resolution swept-frequency reflectometer described by Hollway and Somlo [16] used a 2 m length of waveguide and a swept frequency range from 8·2 GHz to 12·4 GHz, and consequently produced approximately 74 maxima of the interference signal, as shown in Fig. 5.17. The reference reflection discontinuity was a well matched tapered section of waveguide which created an almost constant value of $|\Gamma_r|$ with frequency at the junction of the reduced and normal height waveguides, but it was subsequently proposed that it could be successfully replaced by a steel ball placed at the centre of the broad wall of the waveguide [17]. Such a ball provides an almost constant reflection with frequency, and since it can be held in place by an external magnet it may be readily exchanged with a ball of different diameter to provide a different value for $|\Gamma_r|$. The results in Fig. 5.17 show not only a rapid interference pattern caused by the discontinuity and termination reflections, but also superimposed is a slower interference pattern caused by the frequency responses and interaction of the discontinuity reflection and leakage signals.

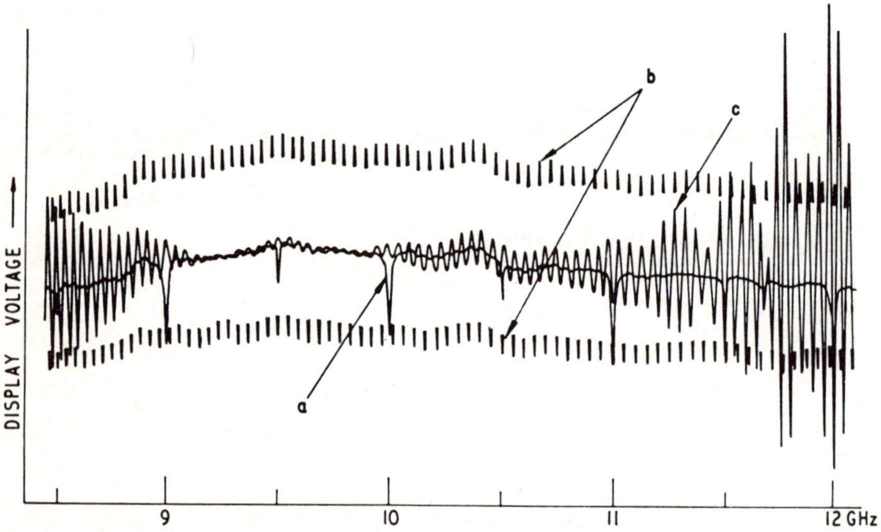

Fig. 5.17. *Typical outputs of the high resolution swept-frequency reflectometer*
Curve a trace (with frequency markers) from a wide band matched load
Curve b extremities of the trace from a calibrating reflection of $|\Gamma_c| = 0{\cdot}05$
Curve c trace from the unknown termination.

[reproduced by permission of IEEE]

Because the electrical path lengths of these two signals are similar, their interference pattern is easily distinguished from that due to the unknown, and therefore the temporary assumption made earlier presents no problems in practice. The results are shown relative to a broadband calibrating termination of $|\Gamma_c| = 0{\cdot}05$, and are for a broadband isolator. Expansion of the region between 9 GHz and 10 GHz is shown in Fig. 5.18, and by comparison with the interference pattern for a matched termination it can be deduced that the isolator had a reflection coefficient of magnitude less than 0·002. When the reference baseline of a reflectionless load is found by sliding a matched load inside the long line and recording the mean value of the trace obtained, the high resolution reflectometer enables a component to be adjusted until $|\Gamma_L| < 0{\cdot}001$.

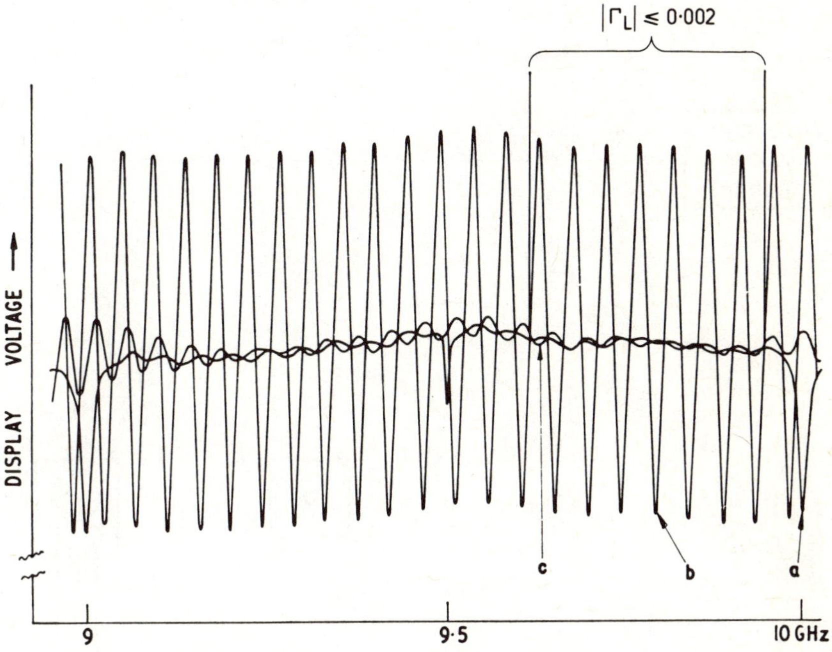

Fig. 5.18. *Expanded view of the 9 GHz to 10 GHz region of Fig. 5.17*
[Reproduced by permission of IEEE]

If port 4 of the coupler is accessible, it may be terminated in a frequency insensitive termination to provide the reference reflection, and then eqn. 5.16 applies directly.

Lacy and Oldfield [18] described a wideband coaxial line reflectometer where the role of the reference reflection is filled by the resistive bridge leakage.

5.4.4 Frequency-swept reflectometers with Smith chart display

A circuit capable of providing a Smith chart display of reflection coefficient over a broad frequency range is shown in Fig. 5.19*a*. A sample of the reflected signal is passed through a broadband quadrature hybrid (3 dB coupler), and the two output signals are each homodyne detected using double balanced mixers as phase detectors, with a sample of the incident signal as the local oscillator. When the

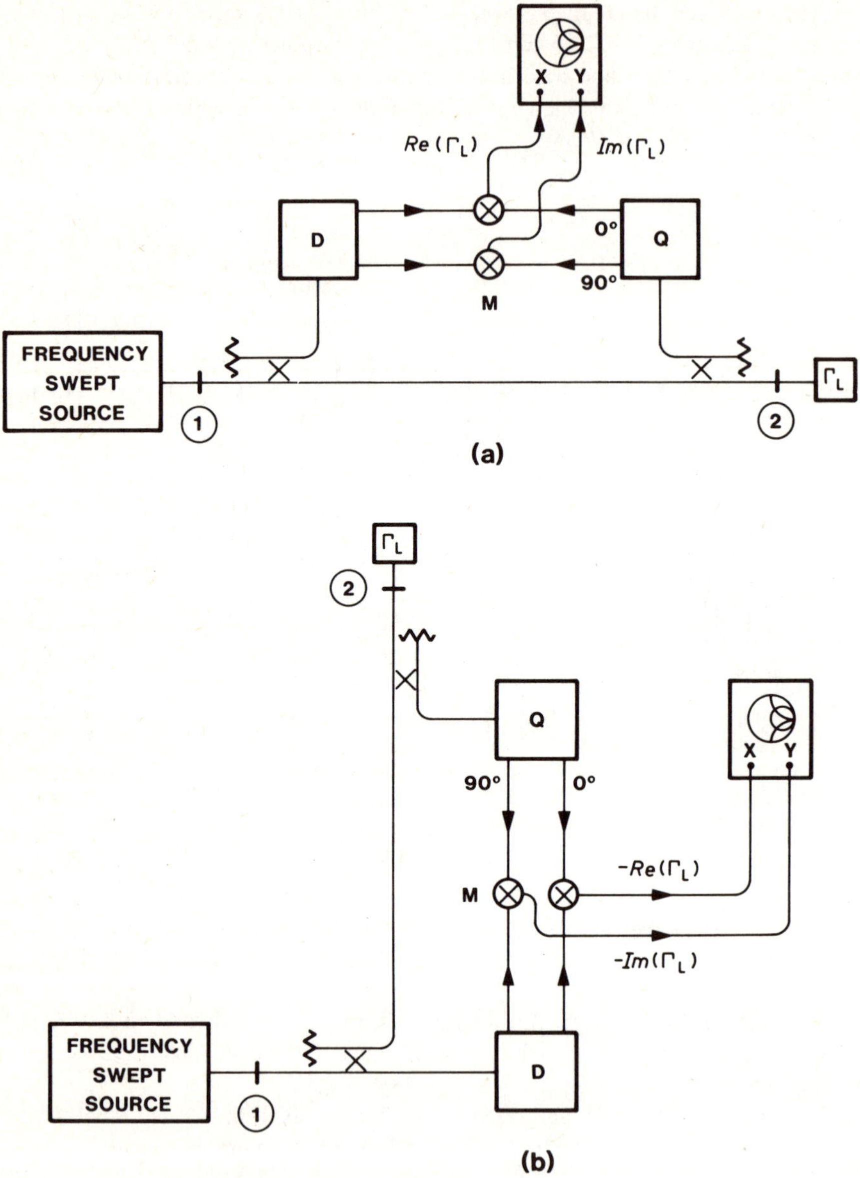

D : DIVIDER

Q : 90° HYBRID (3 dB COUPLER)

M : DOUBLE BALANCED MIXERS

Fig. 5.19. *Swept-frequency reflectometers providing a Smith chart display*
a Circuit using mixers
b Alternative circuit when generator power is limited

electrical path length for the local oscillator signal from port 1 to the mixers is adjusted to be the same as that for the measuring signal from port 1 to the mixers via port 2, the outputs from the mixers are proportional to the real and imaginary parts of Γ_L, and may be connected directly to the X and Y inputs of an oscilloscope to provide a Smith chart display. The use of double balanced mixers is intended to minimize the d.c. offsets associated with the mixer output signals. The values of the main line couplers should be chosen so that the mixers operate at their preferred local oscillator and signal levels, to ensure linear mixing.

The circuit in Fig. 5.19*a* is an extension of the altimeter circuit of Fig. 6.8*a*, and is similar in principle to the locating reflectometer circuit in Fig. 6.10, with the double balanced mixers replacing the magic tee hybrids and their associated detectors and difference amplifiers. As discussed in Section 6.2.2.1, the locating reflectometer also provides a frequency-swept Smith chart display, as does the circuit of Gottfried [19] discussed in Section 4.2.3.1 and shown in Fig. 4.22.

The circuit in Fig. 5.19*b* is an alternative to that in Fig. 5.19*a* when the power available from the generator is a limitation.

References

1 SOMLO, P. I., HOLLWAY, D. L. and MORGAN, I. G.: 'The absolute calibration of periodic microwave phase shifters without a standard phase shifter', *IEEE Trans.*, 1972, **MTT-20**, pp. 552–537

2 PARSHAD, R., CHANDRA, K. and AGARWAL, R. K.: 'A method for accurate measurement of impedance at microwave frequencies using directional couplers in waveguides', *Ind. J. Pure & Appl. Phys.*, 1975, **13**, pp. 325–327

3 THURSTON, W. R.: 'A direct-reading impedance measuring instrument for the UHF range', *Gen. Rad. Exper.*, 1950, **24**

4 SODERMAN, R. A.: 'Improved accuracy and convenience with type 1602-B Admittance meter in VHF and UHF bands', *Gen. Rad. Exper.*, 1953, **28**

5 DOBBIE, L. G. and SOMLO, P. I.: 'Precise absolute technique for measuring low-loss symmetrical linear passive 4-terminal networks at VHF and UHF', *Proc. IEE*, 1963, **110**, pp. 1171–1180

6 WOODS, D.: 'A precision dual bridge for the standardization of admittance at very high frequencies', *Proc. IEE*, 1957, **104/c**, pp. 506–521

7 WILTRON TECH. REVIEW, 1978, 8, Wiltron Co., CA, USA

8 FURUKAWA, S. and AOYAGI, S.: 'Admittance measurements of active elements by a standing-wave ratio meter with a sheilded loop', *Electr. Commun. Japan*, 1969, **52-B**, pp. 65–72

9 ENGEN, G. F. and BEATTY, R. W.: 'Microwave reflectometer techniques', *IRE Trans.*, 1959, **MTT-7**, pp. 351–355

10 ALMASSY, G.: 'A first-order correction to sliding short behaviour with application to the problem of measuring small losses', *IEEE Trans.*, 1971, **IM-20**, pp. 162–169

11 WARNER, F. L.: 'Microwave attenuation measurement (Peter Peregrinus, 1977)

12 SOMLO, P. I.: 'The effect of flange loss on the reflection coefficient of reduced-height waveguide reflection standards', *IEEE Trans.*, 1979, **MTT-27**, pp. 795–797

13 BEATTY, R. W.: 'Magnified and squared VSWR responses for microwave reflection coefficient measurements', *IRE Trans.*, 1959, **MTT-7**, pp. 346–350

14 HOLLWAY, D. L. and SOMLO, P. I.: 'Origin of high-resolution swept-frequency reflectometry', *Microwave Journal*, 1973, **16**, p. 20

15 DE RONDE, F. C.: 'A precise and sensitive reflectometer providing automatic full-band display of reflection coefficient', *IEEE Trans.*, 1965, **MTT-13**, pp. 435–440

16 HOLLWAY, D. L. and SOMLO, P. I.: 'A high-resolution swept-frequency reflectometer', *IEEE Trans.*, 1969, **MTT-17**, pp. 185–188

17 SOMLO, P. I. and HOLLWAY, D. L.: 'Conductive contacting spheres on the centre of the broad wall of rectangular waveguides', *Electron. Lett.*, 1972, **8**, p. 507

18 LACY, P. and OLDFIELD, P.: 'A precision swept reflectometer', *Microwave Journal*, 1973, **16**, p. 20

19 GOTTFRIED, S.: 'A millimeter wave complex reflection coefficient scanner', *IEEE Trans.*, 1973, **IM-22**, pp. 325–330

Chapter 6

Time-domain reflectometry

The measurement methods described in the preceding chapters provide values $\Gamma(f)$ of reflection coefficient as a function of frequency, and fall into the category of frequency-domain reflectometry. The measurement of reflection coefficient at a single frequency provides no information about the physical or electrical location of a discontinuity, and when such information is required the necessary measurements and analysis are carried out by a time-domain reflectometer (TDR). Examples of time-domain reflectometry occur naturally in the echo-location ability of bats and dolphins, and are crudely imitated by man using pulsed radars. These systems, unlike transmission line time-domain reflectometry, must all contend with space loss.

The principle of a transmission line TDR is to send a signal $f(t)$ along the line, and from the character and delay of the reflection (echo) deduce the character and location of the discontinuity. To enable reflections from adjacent discontinuities to be resolved, the rise time of the transmitted signal should be as small as possible. Because the signal $f(t)$ and its frequency spectrum $F(f)$ are related by the Fourier transformations

$$F(f) = \int_{-\infty}^{\infty} f(t)\, e^{-j2\pi ft}\, dt$$
$$f(t) = \int_{-\infty}^{\infty} F(f)\, e^{j2\pi ft}\, df \qquad (6.1)$$

a smaller rise time implies a broader frequency spectrum [1], and therefore a larger system bandwidth is required.

A discontinuity which results in a reflection coefficient $\Gamma(f)$ at the measuring port will produce there a reflected signal $f_r(t)$ given by

$$f_r(t) = \int_{-\infty}^{\infty} \Gamma(f) F(f)\, e^{j2\pi ft}\, df \qquad (6.2)$$

Consequently, discontinuities which are frequency dependent or transmission lines which are dispersive produce an echo which is a distortion of the original transmission, and can be difficult to interpret. As an example, the discontinuity may be sufficiently frequency sensitive that the integral in eqn. 6.2 evaluates to near zero, and hardly any reflection is detected at the measuring port. A TDR is therefore not well suited to measure the characteristics of resonant discontinuities or similar structures.

If a discontinuity in a dispersionless line is frequency independent, and reflects a portion Γ_d of any signal incident upon it, the reflection coefficient seen at the measuring port is

$$\Gamma(f) = \Gamma_d e^{-j2\beta l} = \Gamma_d e^{-j4\pi f l/v} \tag{6.3}$$

where l is the distance to the discontinuity, and v is the velocity of propagation. When this is inserted in eqn. 6.2 it follows from the shift theorem for Fourier transforms that

$$f_r(t) = \Gamma_d f(t-t_l), \quad t_l = \frac{2l}{v} \tag{6.4}$$

In this special case, the echo is a replica of the transmission, and as might be expected, arrives at the measuring port after a delay t_l, from which the effective location of the discontinuity may be deduced. In general, in a dispersionless line, $f_r(t)$ can be interpreted as a function of distance.

A transmission line TDR is particularly well-suited to measure frequency independent discontinuities, and is useful for locating maladjusted connectors, cable cracks, poor transitions, and similar transmission line problems.

With a practical TDR, it is necessary to compromise between the desired resolution and the available bandwidth. With a demand for higher resolution comes the difficulty of generating a signal $f(t)$ with a rise time sufficiently small (bandwidth sufficiently large) to obtain that resolution. In such circumstances, the relationships in eqns. 6.1 and 6.2 can be used to compute the response $f_r(t)$ to a presumed transmission $f(t)$ provided $\Gamma(f)$ is known. For example, the response $f_r(t)$ of a reflecting object to a pulse transmission $f(t) = \delta(t)$, where $\delta(t)$ is the Dirac impulse function, can be determined synthetically from

$$f_r(t) = \int_{-\infty}^{\infty} \Gamma(f)\, e^{j2\pi f t} df \tag{6.5}$$

since the Fourier spectrum of $\delta(t)$ is

$$F(f) = \int_{-\infty}^{\infty} \delta(t)\, e^{-j2\pi f t} dt = 1 \tag{6.6}$$

Similarly, the rise time of an instrument, such as an oscilloscope, can be found by measuring the (transmission) frequency response $T(f)$ of the instrument and computing the time-domain behaviour by substituting $T(f)$ for $\Gamma(f)$ in eqn. 6.5.

Section 6.1 discusses TDR methods which use pulses or steps generated in real time, and Section 6.2 discusses methods which synthesize pulse or step response from the frequency domain response.

6.1 Time-domain reflectometry using sharply rising wavefronts

A simple method of producing a sharply rising wavefront is to impress on the transmission line a voltage pulse or step. The echoes can be displayed on an oscilloscope in real time, and their interpretation is often simple. However, such transmissions are not useful in waveguide systems, since only that part of their spectral energy above the cutoff frequency is transmitted, and that is subject to dispersion.

Time-domain reflectometry uses voltage pulses or steps as the transmitted signal $f(t)$. The rise times must be sufficiently small to enable adjacent discontinuities to be located and distinguished with the desired resolution by timing the arrival of the echoes. Interpreting the reflected signal and locating the discontinuities is simplified if the transmission line responds linearly, is dispersionless, and capable of propagating a broadband signal.

If the transmission line is non-linear, the magnitudes of components of the frequency spectrum of the transmission will be altered not only by the discontinuity, but also by the transmission line. If the line is dispersive, the delay between transmission and reception will vary for different frequency components of the signal, and this differential delay will depend on the distance to the discontinuity. The result of either non-linearity or dispersiveness is to distort the shape of the reflected signal, and, without special processing, the echo may be difficult to interpret.

The more the transmission line or the receiver is bandlimited, the less will be the resolving ability. If the discontinuity characterized by eqn. 6.3 is interrogated by a Dirac impulse, but the system is bandlimited to $B = f_2 - f_1$, then, from eqns. 6.2 and 6.6

$$f_r(t) = 2\Gamma_d B \operatorname{sinc}\left[B\left(t - \frac{2l}{v}\right)\right] \cos\left[\pi(f_1 + f_2)\left(t - \frac{2l}{v}\right)\right] \tag{6.7}$$

where

$$\operatorname{sinc}(x) = \sin(\pi x)/\pi x \tag{6.8}$$

We (somewhat arbitrarily) choose to define the minimum separation r between adjacent discontinuities which still allows them to be resolved, as one half the width of the main lobe of the sinc function. For the dispersionless line, the minimum resolvable separation is therefore

$$r = \frac{v}{2B} \tag{6.9}$$

The rise times of currently available pulse generators are around 50 ps, and therefore video bandwidths of several gigahertz are necessary to fully exploit the potential resolving ability. For this reason, conventional microwave semiconductor detectors are not suitable as receivers, and TDR instruments use sampling methods to reconstruct the echo shape. These methods require the signal to be transmitted repeatedly, and successive echoes are then sampled after slightly different delays. The need to use such methods makes the simple TDR technique expensive to implement.

Fig. 6.1 shows a triangular pulse $f_1(t)$ and a quasi-step function $f_2(t)$ having the same rise time. The respective spectra are [1]

$$\begin{aligned} F_1(f) &= \text{sinc}^2(f) \\ F_2(f) &= -\text{j sinc}(f)/2\pi f \end{aligned} \tag{6.10}$$

which have envelopes which decay at the same rates, and therefore the resolutions obtainable using these waveforms are the same. Because the quasi-step function is somewhat simpler to generate than the triangular or quasi-pulse function, and results in echoes which are more readily interpreted, most currently available TDR instruments use it as the transmission function. Since a rise time of 50 ps corresponds to a propagation distance of about 15 mm in air-dielectric TEM transmission line, individual discontinuities with half this separation can be resolved. Closer discontinuities may still be resolved, but with decreasing definition as the separate echoes increasingly overlap.

There are a number of possible ways of interconnecting the step-function generator, the sampling receiver, and the transmission line or other Device Under Test (DUT) [2, 3]. The most common method is shown in Fig. 6.2. In order to avoid false echoes caused by multiple reflections, the DUT is presented with a matched source impedance either by design of the tee junction, or if the step-function generator is matched, by ensuring that the sampling receiver probe has a high impedance.

6.1.1 *Interpreting a TDR display*

The echoes obtained using a TDR instrument can be analysed using the well known methods applicable to transient phenomena, such as the application of the Heaviside operator or Laplace transformations. In practice, it is often possible to interpret the echoes by referring to a catalogue of waveforms characteristic of simple discontinuities.

As the simplest case, consider a linear, lossless, dispersionless TEM transmission line of real characteristic impedance Z_0 terminated by a resistive load R_L a distance l from the input, as shown in Fig. 6.3*a*. The line is driven by a matched generator which impresses a voltage step of 1 V on the line at port A at time $t = 0$. The voltage step travels along the line with velocity v, and reaches the termination at $t = l/v$. If $R_L = Z_0$, the voltage across the termination rises immediately to 1 V, and no reflected wave is set up. However, if $R_L = 0$, the voltage across the termination is zero, and a reflected voltage step of -1 V propagates back along the transmission line arriving at the input port at $t_l = 2l/v$, and thereafter the voltage there remains at zero. Similarly, if the line is terminated in an open circuit ($R_L = \infty$) a 1 V step will be reflected and after arriving at the input port will cause the voltage there to remain at 2 V indefinitely. These three cases are illustrated in Fig. 6.3*b*. In general, for times $t \geqslant 2l/v$ and resistive loads, the voltage at the input port will

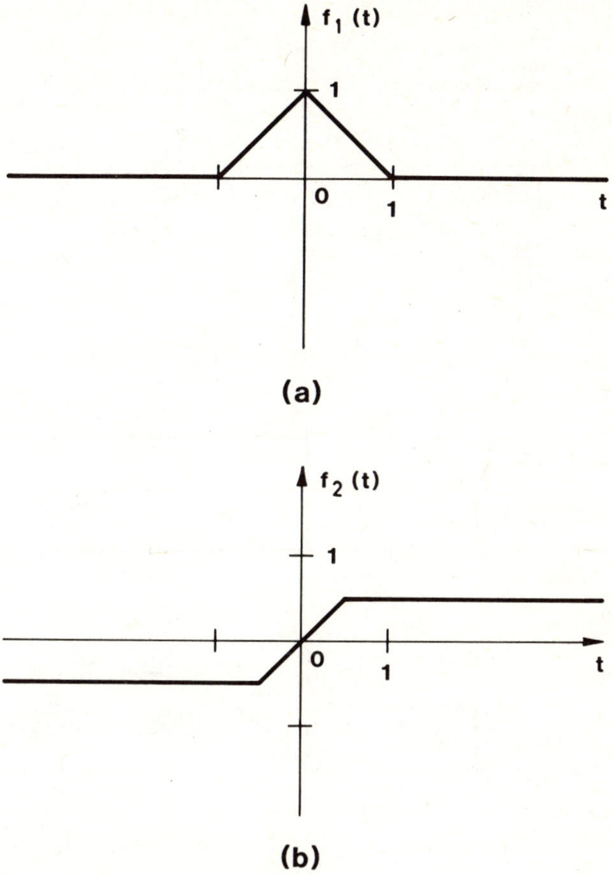

Fig. 6.1. *Approximations to idealized functions*
a Triangular approximation to an impulse
b Ramp-function approximation to a step function

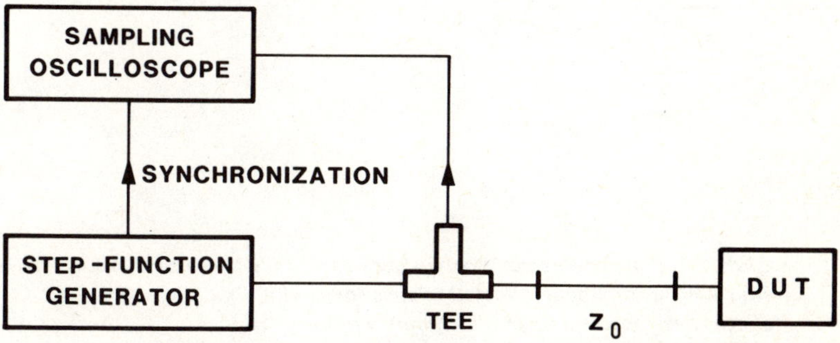

Fig. 6.2. *Simple TDR circuit. To avoid multiple echoes the DUT should be presented with a matched source impedance*

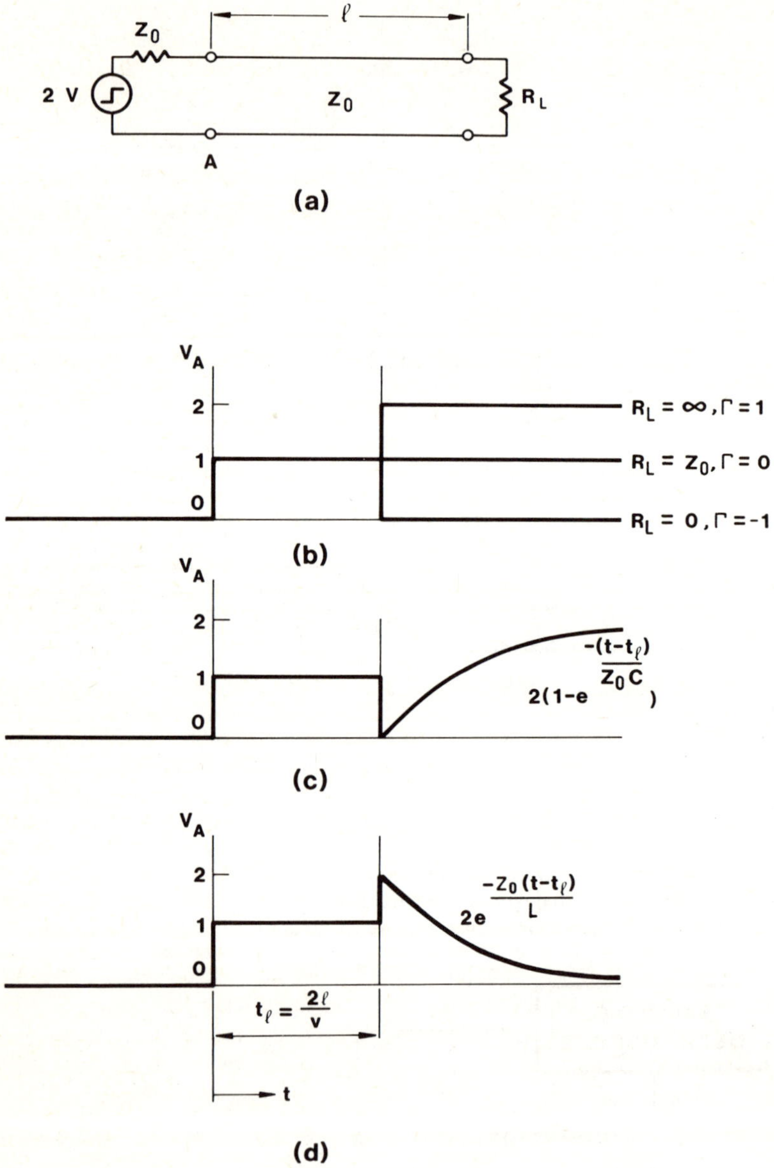

Fig. 6.3. *a Circuit of a matched step function generator driving a lossless, dispersionless, uniform transmission line terminated in a resistive load*
Voltages developed across the line input terminals for
b three values of resistive load
c capacitive load
d inductive load

be $1 + (R_L + Z_0)/(R_L - Z_0) = 1 + \Gamma_L$, and the vertical axis in Fig. 6.3*b* can be scaled linearly in Γ_L.

If the termination is wholly reactive, the well known behaviour of *RL* and *RC* circuits will be evident, with Z_0 acting as the resistor. A terminating capacitor behaves initially as a short circuit, but finally as an open circuit, and leads to the display shown in Fig. 6.3*c*. A terminating inductor behaves initially as an open circuit but finally as a short circuit, and leads to the display shown in Fig. 6.3*d*.

Figs. 6.4*a–f* show the displays expected from various types of discontinuity a distance l along a linear, lossless, dispersionless TEM transmission line. These discontinuities may all be represented by simple two-port networks of zero electrical length. Uniformly distributed series or shunt losses commencing a distance l along the line result in the displays of Figs. 6.4*g* and *h*, respectively. These displays are characterized by linear ramps, which settle at levels corresponding to those of Figs. 6.4*a* and *b*, with *R* and *G* being the total series resistance and shunt conductance, respectively.

6.1.2 Effect of multiple reflections

Fig. 6.5 represents a lossless transmission line with frequency independent discontinuities at distances l_1 and $l_1 + l_2$ from the input. The reflection coefficients associated with these discontinuities are Γ_1 and Γ_2, respectively. A time $t_1 = 2l_1/v$ after a voltage step is impressed on the line, a reflected wave with magnitude $|\Gamma_1|$ will arrive at the input, and be followed after a further delay of $t_2 = 2l_2/v$ by a wave of magnitude $M_2 = |S_{21}||S_{12}||\Gamma_2|$ reflected from the second discontinuity. Because of the multiple reflection between the discontinuities, an infinite succession of reflected waves will then appear at intervals of t_2, the *n*th in the succession having a magnitude $M_2 r^n$, where $r = |S_{22}||\Gamma_2|$.

For many practical applications of a TDR, individual reflection coefficients are small, and the succession of echoes or 'ghosts' decays rapidly. For example, with $|S_{22}| = |\Gamma_2| = 0{\cdot}2$, each ghost has a magnitude only 4% of its predecessor. If, for example, a TDR is being used to monitor the adjustment of a broadband matching tuner, the ghosts become insignificant well before the final adjustment.

A second consequence of multiple discontinuities is that reflections from further along the line are reduced because of energy lost or reflected from preceding discontinuities. In Fig. 6.5 a wave of unit magnitude is reduced to magnitude $|S_{21}|$ after passing the first discontinuity. If this discontinuity is lossless and reciprocal, the TDR display of the reflection from the second discontinuity must be scaled by the factor $1/(1 - |\Gamma_1|^2)$. If $|\Gamma_1|$ is small, this factor is not significant, but if, for example, $|\Gamma_1| = 0{\cdot}5$, the TDR display of $|\Gamma_2|$ should be increased by 33%. Of course, if the first discontinuity is lossy, the display of subsequent reflections must be appropriately scaled.

It is interesting to imagine the voltage step replaced by a continuous wave of a single frequency. The successive reflections, correctly phased, may be combined mathematically as the sum of a geometric series, and the reflection coefficient Γ_{in} seen at the input to the line is

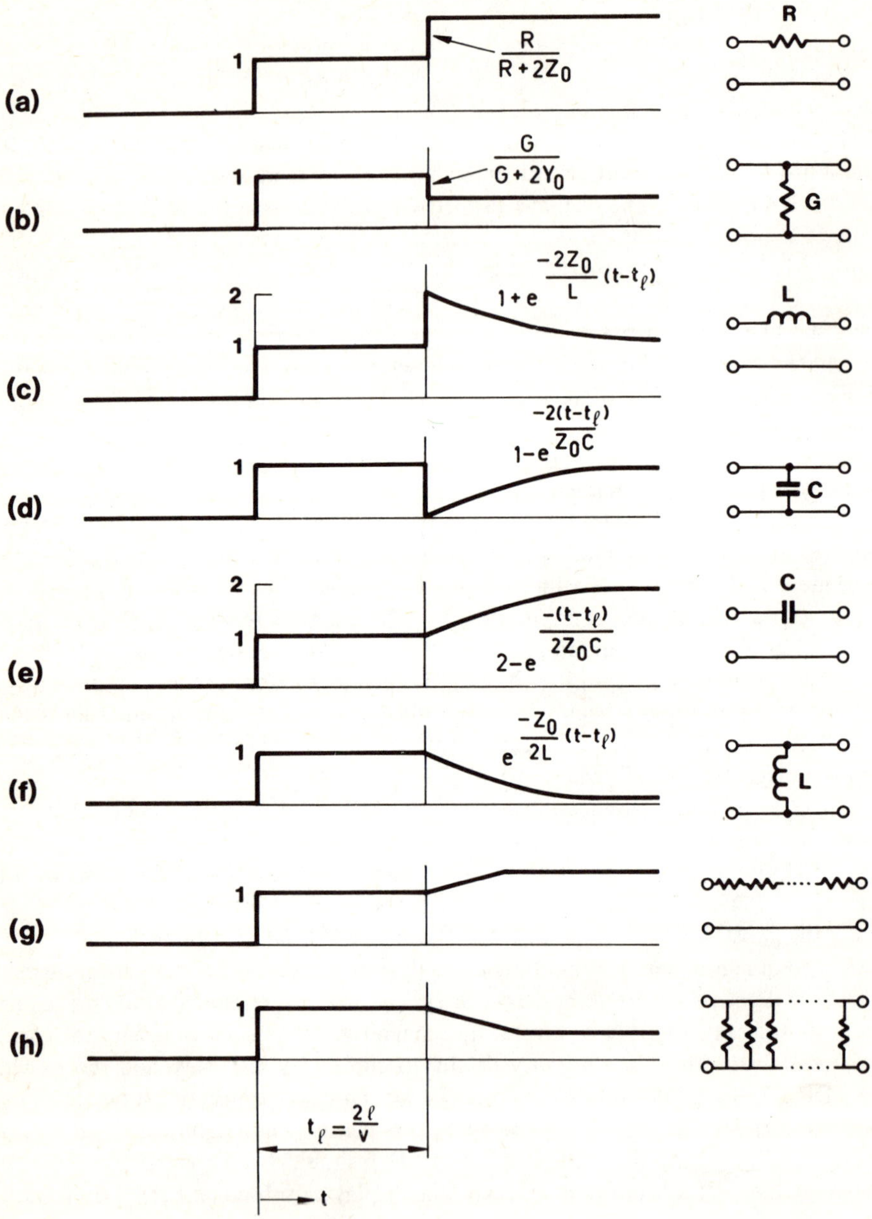

Fig. 6.4. *Idealized TDR responses from discontinuities at distance ℓ along a lossless, dispersionless, uniform transmission line*
a–f Various lumped discontinuities
g–h Distributed losses starting at distance ℓ

$$\Gamma_{in} = S_{11} + \frac{\Gamma_2 S_{12} S_{21} e^{-j2\theta}}{1 - \Gamma_2 S_{22} e^{-j2\theta}} \tag{6.11}$$

This is identical to the result obtained by applying flow graph analysis to Fig. 6.5, and illustrates that a well known result in the frequency domain is obtainable from time-domain considerations.

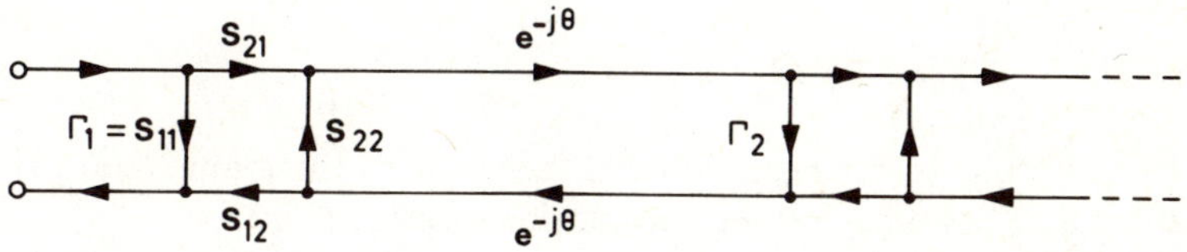

Fig. 6.5. *Flow graph of two discontinuities separated by an electrical distance $\theta = \beta \ell_2$ along a lossless transmission line*

6.1.3 Detection of small reactive discontinuities

It has been assumed in preceding sections that time constants associated with the discontinuity are appreciably larger than the response time of the TDR system. However, smaller time constants can be measured provided that system noise and time-jitter levels are sufficiently low [4].

Consider a small inductor L located a distance l from the input of a dispersionless transmission line of real characteristic impedance Z_0, and in series with the line. The reflection coefficient at the input port is

$$\Gamma(f) = \frac{j\pi f L}{Z_0 + j\pi f L} \; e^{-j4\pi f l/v} \tag{6.12}$$

which may be approximated by

$$\Gamma(f) \simeq j\,\frac{\pi f L}{Z_0} \; e^{-j4\pi f l/v} \tag{6.13}$$

provided L is sufficiently small that within the system bandwidth $\pi f L \ll Z_0$. When this is inserted into eqn. 6.2, use of the Fourier transform shift theorem and eqn. 6.1 show that

$$f_r(t) = \frac{L}{2Z_0} \, f'(t - t_l) \tag{6.14}$$

The reflected wave is therefore the time derivative of the transmitted signal, delayed by $t_l = 2l/v$ and weighted by $L/2Z_0$. The TDR display for a voltage step transmission is shown in Fig. 6.6, from which both $f'(t - t_l)$ and the maximum value of $f_r(t)$ may be found, and hence L/Z_0. A similar display is obtained if a small capacitor C shunts the transmission line, but in this case

$$f_r(t) = \frac{-CZ_0}{2} \, f'(t - t_l) \tag{6.15}$$

When a small capacitor is in series with the line, or a small inductor shunts the line, it is difficult to distinguish the reflection from that created by an open circuit or short circuit, respectively. However, the signal $f_t(t)$ which passes the discontinuity and appears across the (matched) termination is, for a small series capacitor C

$$f_t(t) = 2CZ_0 f'(t - t_L) \tag{6.16}$$

and for a small shunt inductor L

$$f_t(t) = \frac{2L}{Z_0} f'(t - t_L) \tag{6.17}$$

where t_L is the time of propagation from the input to the termination [5]. These may be categorized as time-domain transmission measurements.

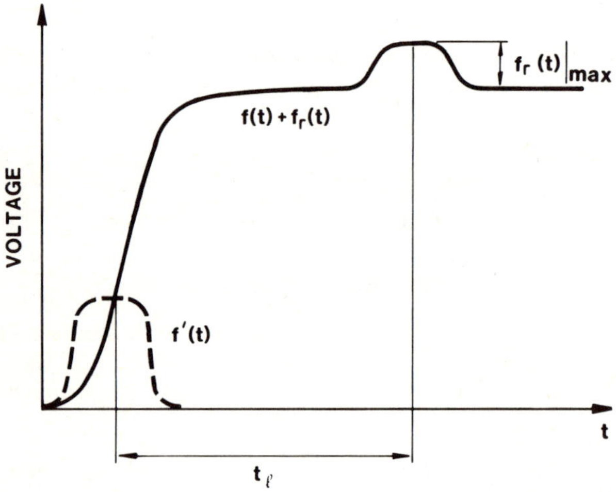

Fig. 6.6. *TDR display for a small series inductive discontinuity*

An alternative method for finding the time constant of a small reactive discontinuity is to measure the area under the display of $f_r(t)$ or $f_t(t)$. It follows immediately from eqns. 6.16 and 6.17 that this area depends simply on the time constant and the final value of $f(t)$. Consequently, if $f(t)$ is a voltage step, the shape of the step is unimportant.

6.1.4 Techniques to improve sensitivity

The possible methods by which TDR sensitivity can be improved have been summarized by Cowan [6] and include averaging, differencing, smoothing, and autocorrelating.

Elliot [7, 8] improved the sensitivity of a TDR display by first averaging a portion of the display for a specified period, and then averaging the same 'time window' when a relatively long, plain, uniform terminated transmission line was substituted for the DUT. The averaging increased the signal-to-noise ratio, and by differencing the two results smaller reflections from the DUT could be recognized in spite of the presence of internal reflections in the measuring instrument. Elliot claims a 200–500 times increase in sensitivity over a simple TDR circuit. The removal of the background clutter by differencing is similar to the method which was used by Hollway [9] and is discussed in Section 6.2.1.

6.1.5 Spectral weighting to improve axial resolution

When a TDR system is bandlimited, the axial resolution achieved may be increased somewhat, at the expense of signal-to-noise ratio, by spectral weighting techniques. Appropriately weighting the propagating spectral components of $f(t)$ leads to 'super resolution', and is discussed by Chambers [10]. Application of related deconvolution techniques to synthetic pulse measurements is discussed in detail in Section 6.2.3.

6.1.6 Operations in dispersive lines

It has already been pointed out that the use of a voltage pulse or step for TDR measurements in a waveguide system is inefficient in that much of the energy in the signal does not propagate. Because of dispersion, the signals at frequencies which do propagate do so at different velocities, and a reflection which is characteristic of a particular type of discontinuity in a dispersionless line returns distorted. To overcome these limitations, a compromise between bandwidth and resolving ability can be arranged by transmitting a signal similar to that of conventional radar. The signal is pulse modulated CW with an effective bandwidth sufficiently narrow that dispersive effects are not serious. Various systems are available commercially, some operating with a fixed carrier frequency, and others tuneable over a wide range (in one case from 1 GHz to 18 GHz). Pulse widths, which may be as small as 2 ns with rise times of about 1·5 ns, limit axial resolution to around 30 cm in waveguide, but are adequate to indicate the approximate location of a discontinuity in a long waveguide run.

Mackay and Penstone [11] describe a laboratory-type narrowband TDR which uses a carrier frequency of 70 MHz with a pulse width of 80 ns. Because the pulse duration incorporates only a few cycles of the carrier frequency, envelope detection is difficult, and a scheme of synchronous detection is used. This offers the possibility of detecting both the in-phase and quadrature components of the reflection, and thereby determining both the magnitude and phase of the reflection at the discontinuity. The detection of the quadrature component is achieved after introducing an extra $\lambda/4$ length of transmission line into the path of the reference signal. By using a computer to subtract from the reflected signal one measured previously, very small changes in the line can be revealed. The high sensitivity in this differential mode permits the detection of reflecting objects placed near a leaky cable.

6.1.7 Network analysis using time-domain reflectometry

Just as the Fourier transform relationships in eqn. 6.1 allow the pulse response of a system to be synthesized from the frequency response, they allow the frequency response to be calculated from the pulse response. This latter approach to network analysis is attractive in principle, because a single pulse response measurement enables the response at all frequencies to be computed. Of course, the frequency components in the pulse response must be normalized to the same components in the transmitted signal, and the results are then, in principle, independent of the pulse shape. The absence of some spectral components in the transmitted signal leaves the network response undefined at these frequencies, and the accuracy which can be obtained with small spectral components is limited by the signal-to-noise ratio.

Nicolson *et al.* [12] used this technique after connecting a computer to the output of a sampling oscilloscope. The computer applied signal averaging techniques, and performed a fast-Fourier transformation to compute the frequency domain information. Measurements were performed on both closed and open structures. The great advantage of using TDR methods on open structures is that unwanted echoes are separated in time and may be discarded. Open structures, such as antennas, may be tested in an ordinary laboratory, and anechoic chambers or open-air ranges are not needed.

Andrews and Gans [13] describe an automatic network analyser using time-domain observations with digital Fourier transformation. Conventional and TDR-based attenuation measurements on 10 dB and 40 dB attenuators using a spectrum extending up to 12·5 GHz were claimed to agree to within 1·5%.

A low-cost time-domain automated network analyser was described by Rigg *et al.* [14] who tested the frequency responses of low-pass and high-pass filters.

6.1.8 Industrial applications

Apart from the common uses of reflectometry to locate faults on transmission lines, geological anomalies (such as buried pipes), and structural anomalies (such as reinforcing rods), many other industrial uses have been found. Two examples are given here.

Cormack [15] measured the thickness of silver plating on 50–50 Ni–Fe alloy wires by making a short test sample ($<$ 3 cm) the inner of a short-circuited coaxial line. This provided a series L–R combination, and the height of the TDR display could be correlated with the wire conductivity.

Kent [16] found the dielectric properties of frozen fish after computing the frequency response from the TDR result. The complex dielectric constant which provided the same values of reflection coefficient was found by iteration.

6.2 Synthetic pulse time-domain reflectometry and the locating function

Synthetic pulse reflectometry is based on eqn. 6.2, which shows how to synthesize the time-domain response $f_r(t)$ of a system from the frequency domain response

$\Gamma(f)$. Whether the system be dispersive or dispersionless, eqn. 6.2 will synthesize the time-domain function which would have been obtained if a real-time signal $f(t)$ had been transmitted. However, because synthetic pulse reflectometry requires signal processing, be it analogue or digital, the opportunity exists to remove the distorting effects of dispersion, and construct the time-domain response of the same discontinuities in a dispersionless system. When the purpose of pulse measurements is to locate discontinuities, this second processing mode is preferable. We must therefore distinguish between true synthetic pulse time-domain reflectometry, which constructs $f_r(t)$ from the frequency response of the system, and the locating function, which is the result of processing designed to locate discontinuities in either dispersive or dispersionless systems.

Just as $\Gamma(f)$ is a complex reflected signal as a function of frequency, we can define the locating function $\Gamma'(l)$ to be a complex reflected signal as a function of distance along the transmission line. In general

$$\Gamma'(l) = \frac{1}{\nu_2 - \nu_1} \int_{\nu_1}^{\nu_2} G(\nu)\, e^{j4\pi\nu l} d\nu \tag{6.18}$$

$G(\nu)$ is the reflection coefficient measured as a function of the wavenumber ν at the input port, and

$$G(\nu) = G[f/v_p(f)] = \Gamma(f) \tag{6.19}$$

where $v_p(f)$ is the phase velocity and is, in general, a function of frequency. An equivalent definition of the locating function is therefore

$$\Gamma'(l) = \frac{1}{\nu_2 - \nu_1} \int_{f_1}^{f_2} W(f)\Gamma(f)\, e^{j4\pi f l/v_p(f)} df \tag{6.20}$$

where

$$W(f) = \frac{d\nu}{df} = \frac{v_p(f) - f\, d/df\, [v_p(f)]}{[v_p(f)]^2} \tag{6.21}$$

In particular, for waveguide systems where, from eqns. 1.39, 1.40, 1.46

$$v_p(f) = \frac{\omega}{\beta} = \frac{v}{[1 - (f_c/f)^2]^{1/2}} \tag{6.22}$$

eqn. 6.20 becomes

$$\Gamma'(l) = \frac{1}{v[\nu_2 - \nu_1]} \int_{f_1}^{f_2} \frac{\Gamma(f)\, e^{j4\pi f l/v_p(f)}}{[1 - (f_c/f)^2]^{1/2}} df \tag{6.23}$$

A frequency independent discontinuity with reflection coefficient Γ_l at $l = l_1$ produces, at the measuring port,

$$G(\nu) = \Gamma_l e^{-j4\pi\nu l_1} \tag{6.24}$$

and from eqn. 6.18

$$\Gamma'(l) = \Gamma_l \operatorname{sinc}\,[2(\nu_2 - \nu_1)(l - l_1)]\,\mathrm{e}^{\mathrm{j}2\pi(\nu_1 + \nu_2)(l - l_1)} \tag{6.25}$$

Notice in particular that when $|\Gamma'(l)|$ is a maximum, at $l = l_1$, then

$$\Gamma'(l_1) = \Gamma_l \tag{6.26}$$

and both the magnitude and phase of the discontinuity reflection coefficient can be found.

Using the same definition for minimum resolvable separation r as in Section 6.1

$$r = \frac{1}{2(\nu_2 - \nu_1)} \tag{6.27}$$

which for a dispersionless line with $v_p(f)$ having a constant value v gives the same result as in eqn. 6.9. However, for dispersive systems in which the wavenumber is not directly proportional to frequency, the minimum resolvable separation does not depend simply on B. For waveguide systems with phase velocities given by eqn. 6.22

$$r = \frac{v}{2[\sqrt{(f_2^2 - f_c^2)} - \sqrt{(f_1^2 - f_c^2)}]} < \frac{v}{2B} \tag{6.28}$$

Consequently, not only is the resolving ability greater (minimum separation less) in a dispersive waveguide system than in a dispersionless system using the same bandwidth, but it also increases as the lower frequency limit tends toward cutoff.

If the line is dispersionless, eqn. 6.20 simplifies to

$$\Gamma'(l) = \frac{1}{B}\int_{f_1}^{f_2} \Gamma(f)\,\mathrm{e}^{\mathrm{j}4\pi f l/v}\,\mathrm{d}f \tag{6.29}$$

and, not surprisingly, $\Gamma'(l)$ is intimately related to the response $f_r(t)$ of the band-limited system to a Dirac impulse, by

$$f_r\left(\frac{2l}{v}\right) = B[\Gamma'(l) + \Gamma'^*(l)] \tag{6.30}$$

Unlike real pulse reflectometry, synthetic pulse reflectometry and the locating function do not provide an immediate display of the location of discontinuities. Apart from the time taken to measure $G(\nu)$ or $\Gamma(f)$, an integral must be evaluated for each distance l of interest.

The integral may be evaluated either digitally or by analogue means. Digital methods have the potential advantage of high accuracy, but require the data to be stored for subsequent processing. The processing is performed after approximating the integral by a summation over a number of discrete values of the variable in the range of integration. Examples of TDR systems using digital processing are discussed in Section 6.2.1.

If the integral is to be evaluated by analogue means, the integration can be performed continuously over the frequency range. Examples of synthetic TDR instruments using analogue processing are discussed in Sections 6.2.3 and 6.2.4. In principle, it is not necessary to transmit the frequency components sequentially. Some reflectometers are operated by transmitting a number of discrete frequency components simultaneously, and indeed could be operated with white noise.

Like real pulse reflectometry, the resolving ability of synthetic pulse reflectometry is restricted by the practical limitation of finite bandwidth. However, apart from the opportunity to remove the effects of dispersion, synthetic reflectometry has advantages over real pulse reflectometry, including:

(*a*) The full output power of the oscillator is available for all spectral components, and the synthesized reflection is therefore equivalent to that from a very powerful pulse.

(*b*) Because no small rise time wavefronts must be generated or detected, the electronic circuitry required can be simple, and a sampling oscilloscope is not required.

(*c*) The upper frequency limit is not restricted by state-of-art rise times, but only by the frequency of available oscillators.

(*d*) The bandwidth of the synthetic pulse is known exactly.

(*e*) In high-pass systems (for example, waveguides with a cutoff frequency), all of the transmitted energy can be directed into the passband. In the case of a real-time TDR, much of the pulse energy is filtered.

(*f*) As the centre frequency of a synthetic pulse is increased, it is often possible and practical to maintain constant the relative bandwidth, defined by $2(f_2 - f_1)/(f_2 + f_1)$. As a result, the absolute resolving ability of the synthetic pulse TDR in a dispersionless system will improve proportionally with the centre frequency. A similar improvement is achieved in a dispersive system if $2(\nu_2 - \nu_1)/(\nu_2 + \nu_1)$ is maintained constant.

If the signals passing through a system are measured instead of those reflected, an expression similar to eqn. 6.2 can be used to synthesize the transfer response of the system to a pulse input.

6.2.1 Locating reflectometry using digital processing

The high resolution swept-frequency reflectometer described in Section 5.4.3 produces a reflection coefficient at the input port which is a record of the interference between the reference and unknown discontinuities, and it follows from eqn. 5.31 that the period of this interference pattern depends on the discontinuity separation. Hollway [9] exploited this effect in the comparison reflectometer shown in Fig. 6.7. If the reflection coefficient $G(\nu)$ of the device under test is small, then in the presence of a reference discontinuity at the input to the line it can be shown that the output $S(\nu)$ of the square-law detector is approximately proportional to

$$G(\nu) + K + G^*(\nu) \qquad (6.31)$$

where K is a constant. $S(\nu)$ is recorded digitally over the wavenumber range from ν_1 to ν_2, and subsequently processed according to

$$\frac{1}{\nu_2-\nu_1}\int_{\nu_1}^{\nu_2} S(\nu)\,e^{j4\pi\nu l}\,d\nu \tag{6.32}$$

which, from comparison with eqn. 6.18, is proportional to

$$\Gamma'(l) + K\,\text{sinc}\,[2l(\nu_2-\nu_1)]\,e^{j2\pi l(\nu_1+\nu_2)} + \Gamma'^*(-l) \tag{6.33}$$

The comparison reflectometer thus determines the locating function without the need to measure phases.

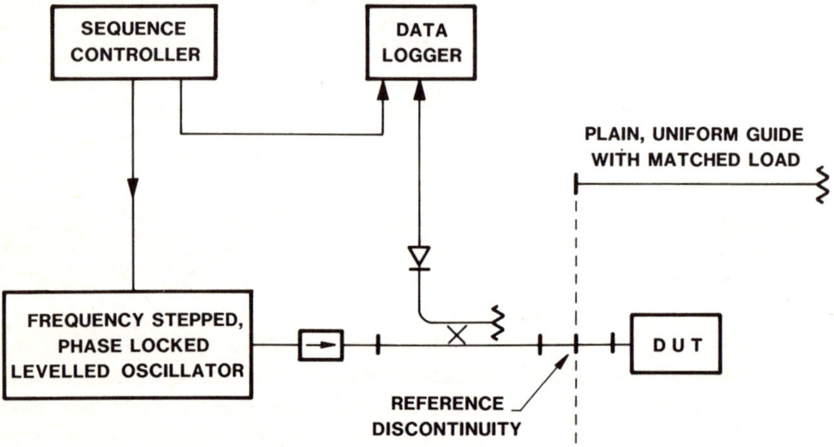

Fig. 6.7. *Comparison reflectometer system of Hollway [9]. The digital Fourier transform of the logged data provides the axial distribution of reflections in the DUT*

Hollway introduced a length of plain transmission line between the reference discontinuity at $l = 0$ and the device under test in order to separate the locating function $\Gamma'(l)$ from the other responses evident in eqn. 6.33.

Imperfections in the measuring apparatus can cause the locating function to be noisy, but much of this noise can be removed by subtracting the result obtained when a plain section of transmission line terminated in a match is connected to the measuring port. This subtraction can be done most efficiently for each value of ν before computing eqn. 6.32. Using this method, Hollway showed that the comparison reflectometer had a residual background noise level equivalent to a reflection coefficient of 0·00006 or less in X-band waveguide, and was therefore capable of detecting very small discontinuities. A cosine squared weighting was introduced into eqn. 6.32 to reduce the sidelobe levels in the locating function (at the expense of resolving ability), and single reflections could be located with an uncertainty of about ± 0·3 mm.

The synthetic pulse technique has also been applied in free-space by Robinson, Weir and Young [17] using horn antennas to find the time-domain behaviour of reflecting plates, spheres, and stonework. Hines and Stinehelfer [18] used a microwave network analyser to perform the frequency-domain measurements, and by appropriately weighting the spectral components, computed the time-domain response of several dispersionless networks to synthetic impulses, pulses, and step functions.

6.2.2 *Locating reflectometry using analogue processing*

The use of a linearly frequency modulated signal to locate discontinuities is probably best known in association with airborne altimeters [19], but is also useful for locating discontinuities in dispersionless transmission lines.

The basic diagram of a dispersionless transmission line discontinuity location system is shown in Fig. 6.8. The transmitted signal sweeps linearly over a bandwidth B in time t_s, and is partially reflected by a discontinuity at distance l_1. Because of the time delay before the reflected signal reaches the mixer, the two signals there differ in frequency by

$$f_l = \frac{2B}{vt_s} l_1 \tag{6.34}$$

where v is the velocity of propagation. The frequency spectrum of the video signal is, from eqn. 6.1, proportional to

$$F(f) = \int_0^{t_d} e^{-j2\pi(f-f_l)t} \, dt \tag{6.35}$$

with $t_d = t_s - 2l_1/v$, and may be written in terms of a variable l as

$$F\left(\frac{2B}{vt_s} l\right) = \int_0^{t_d} e^{-j4\pi B(l-l_1)t/vt_s} \, dt \tag{6.36}$$

By comparison with eqn. 6.29, the above expression describes a locating function, and since for most practical transmission systems $t_d \simeq t_s$, the magnitude of the frequency spectrum of the video signal may be interpreted as

$$a|F(f)| = |\Gamma'(l)| = \left| \text{sinc} \left[\frac{2B}{v} (l-l_1)\right] \right| \tag{6.37}$$

where a is a normalization constant. A linear FM transmission therefore has the minimum resolvable separation in eqn. 6.9, which is independent of sweep rate.

Yamaura and Hidaka [20] described a double-frequency-swept locating reflectometer in which both the frequency and the sweep duration are swept. When the difference frequency in eqn. 6.34 is bandpass filtered, the sweep time t_s necessary to produce the maximum output from the filter is directly proportional to the distance l_1 to the discontinuity, and a plot of the output against t_s can be interpreted as $|\Gamma'(l)|$ with the minimum resolvable separation of eqn. 6.9.

The digital processing for the comparison reflectometer described in the preceding section was replaced by analogue processing by Somlo and Hollway [21] using the circuitry shown in Fig. 6.9. The operation of this system will be described for a frequency independent discontinuity a distance l_1 from the reference discontinuity, and it will be assumed initially that the signal is swept linearly in wavenumber.

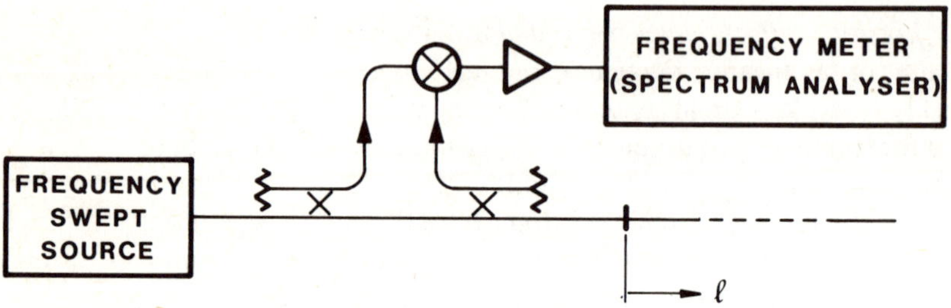

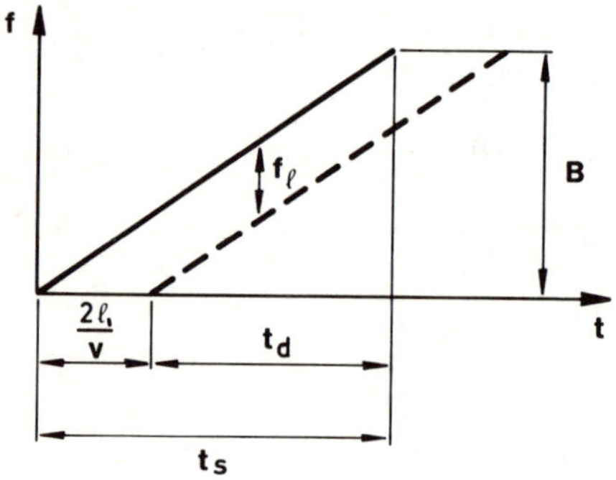

Fig. 6.8. *Swept-frequency discontinuity locating system, also used as the FM-radar altimeter*
a Basic circuit
b ——— transmitted signal
– – – reflected signal at the plane $\ell = 0$

As the signal frequency is swept, the combined $G(\nu)$ components in eqn. 6.31 oscillate with a constant frequency, proportional to l_1. Simultaneously, the VSWR pattern in the second arm passes beneath the probe, and results in the probe detector having an output of a constant frequency which depends on the distance l between the probe and the short circuit. When the two frequencies are identical,

$l = l_1$. A phase-sensitive detector, referenced to the probe detector output in the second arm, acts as a tuned narrowband receiver, and compares the signals from the two arms. After integration, a signal proportional to the real part of $\Gamma'(l)$ is obtained. The imaginary part of $\Gamma'(l)$ can be generated by identical processing provided that a phase shift of 90° is introduced into the path of one of the signals. $|\Gamma'(l)|$ can be plotted using an X–Y recorder. If the x-axis traverse is used to pull the sliding short circuit, the plot has a 1:1 distance scale, and the component under test can later be laid on to the recording and the location of any discontinuities seen directly.

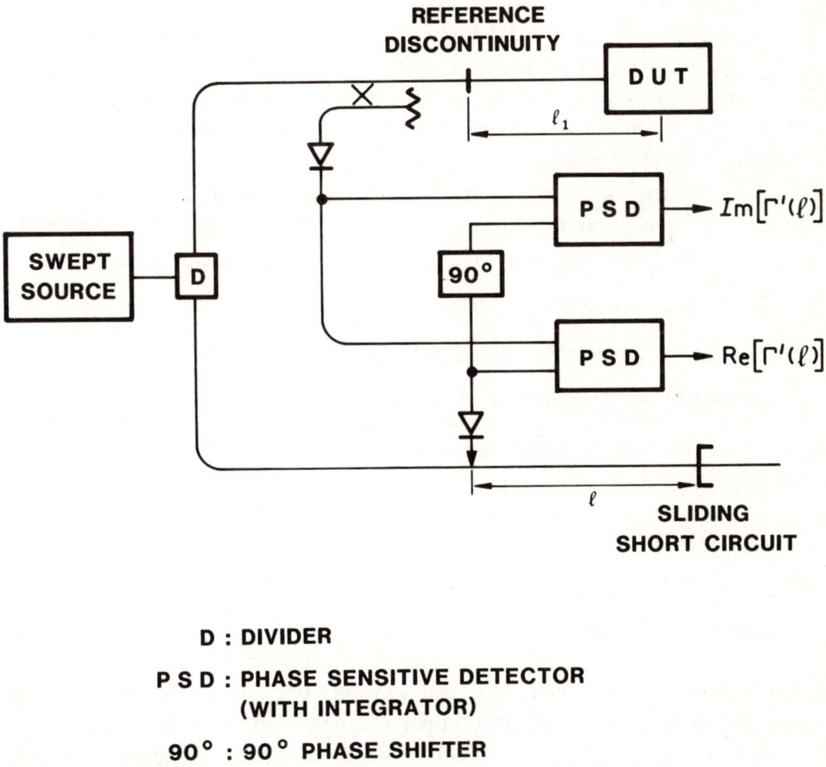

Fig. 6.9. *Locating reflectometer of Somlo and Hollway [21]. The necessary phase detection is performed at video frequencies, and is followed by integration*

In their implementation, Somlo and Hollway [21] swept the signal linearly in frequency rather than wavenumber. Although this leads to a distorted output from a dispersive system, it is shown in the next section that the resolving ability of the reflectometer is not affected significantly by the amount of dispersion normally encountered in waveguides.

Disadvantages of this reflectometer are the need for relatively complicated circuitry, and the restriction on the sweep rate imposed by the range of video frequencies which can be processed. A variant of the circuit in Fig. 6.9 was given by Detlefsen [22] who obtained the reference signal by movement of a slotted line type probe.

6.2.2.1 Locating reflectometer: The locating reflectometer is an instrument incorporating analogue processing, and was first described by Somlo and Hollway [21] and then later in greater detail by Somlo [23]. The instrument is capable of producing the locating function $\Gamma'(l)$ and the reflection coefficient $\Gamma(f)$ of the DUT, or its pulse response.

The circuit is shown in Fig. 6.10, where, for convenience, any signal proportional to the reflection $G(\nu)$ (or $\Gamma(f)$) from the DUT is denoted by a vector A. The vector B denotes any signal proportional to the reference reflection $\Gamma_l \exp(-\mathrm{j}4\pi\nu l)$ from the frequency independent reflection Γ_l at location l in the second arm of the reflectometer. Signal A is separated into equal parts by divider D, while signal B is split into quadrature components by the hybrid Q. Signals A and B are passed through a magic tee H, and their sum and difference are square-law detected and combined in a difference amplifier. The amplifier output is proportional to

$$|A+B|^2 - |A-B|^2 = 4\,\mathrm{Re}\,(AB^*) \tag{6.38}$$

Similarly, the output from processing signals A and $\mathrm{j}B$ is

$$|A+\mathrm{j}B|^2 - |A-\mathrm{j}B|^2 = 4\,\mathrm{Im}\,(AB^*) \tag{6.39}$$

Each of these outputs is integrated over the duration t_s of each sweep of the repetitively swept source, resulting in

$$\mathrm{Re}\int_0^{t_s} G(\nu)\,\mathrm{e}^{\mathrm{j}4\pi\nu l}\,\mathrm{d}t \tag{6.40}$$

and

$$\mathrm{Im}\int_0^{t_s} G(\nu)\,\mathrm{e}^{\mathrm{j}4\pi\nu l}\,\mathrm{d}t \tag{6.41}$$

When the source is swept such that the wavenumber changes linearly with time, the integrator outputs are seen, by comparison with eqn. 6.18, to be proportional to the real and imaginary parts of $\Gamma'(l)$, respectively, and may be used to create a display of the locating function as l is varied.

The locating reflectometer has several advantages over the comparison reflectometer. The outputs of the difference amplifiers, as given by eqns. 6.38 and 6.39 are, with $l = 0$, proportional to

$$\mathrm{Re}\,[G(\nu)] = \mathrm{Re}\,[\Gamma(f)]\,; \quad \mathrm{Im}\,[G(\nu)] = \mathrm{Im}\,[\Gamma(f)] \tag{6.42}$$

and consequently can be connected directly to the X and Y inputs of a CRO to provide a Smith chart display of the reflection coefficient $\Gamma(f)$ as a function of frequency. Simple electronic circuitry is possible because of the use of a hybrid junction to produce a 90° phase shift over the swept range. The locating reflectometer is operable over a wide range of sweep rates, and because it has no reference discontinuity in the measurement arm it is not subject to the approximations necessary to derive eqn. 6.31 for the comparison reflectometer. If the reflectometer is used at a fixed frequency, the Smith chart display is extremely convenient for tuning components to be a match, as discussed in Section 3.3.2.7.

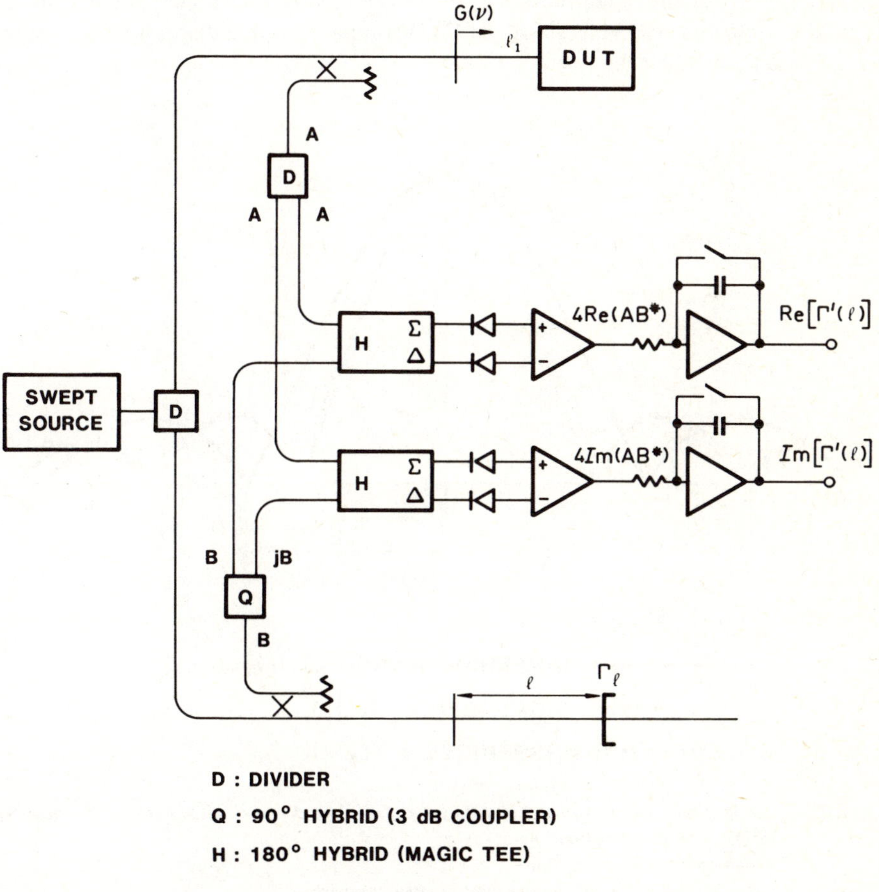

Fig. 6.10. *Locating reflectometer of Somlo [23]. The necessary phase detection is performed at microwave frequencies using hybrid junctions*

In practice, it is easier to sweep frequency linearly with time, in which case the outputs of the integrators are proportional to the real and imaginary parts of $\Gamma'_a(l)$ where

$$\Gamma'_a(l) = \frac{1}{\nu_2 - \nu_1}\int_{f_1}^{f_2} \Gamma(f)\, e^{j\, 4\pi f l / v_p(f)}\, df \tag{6.43}$$

which is termed the 'approximate locating function', and differs from $\Gamma'(l)$ in eqn. 6.20 by the absence of the factor $W(f)$ in the integrand. For dispersionless lines, $W(f) = 1$ and $\Gamma'_a(l) = \Gamma'(l)$. Fig. 6.11 compares $|\Gamma'_a(l)|$ with $|\Gamma'(l)|$ for a frequency independent discontinuity at $l = l_1$, and the reflectometer operating from 8 GHz to 12 GHz. For a dispersionless line, $r\,(\mathrm{mm}) \simeq 38\,v/c$ from eqn. 6.27, but for X-band waveguide with r given by eqn. 6.28 and $f_c = 6{\cdot}56\,\mathrm{GHz}$, $r\,(\mathrm{mm}) \simeq 27\,v/c$. In the waveguide $|\Gamma'_a(l)|$ is seen to have a slightly wider main lobe than $|\Gamma'(l)|$, and is not the magnitude of a sinc function, but nevertheless retains almost all of the improved resolving ability of $\Gamma'(l)$ compared with a dispersionless system using the same bandwidth.

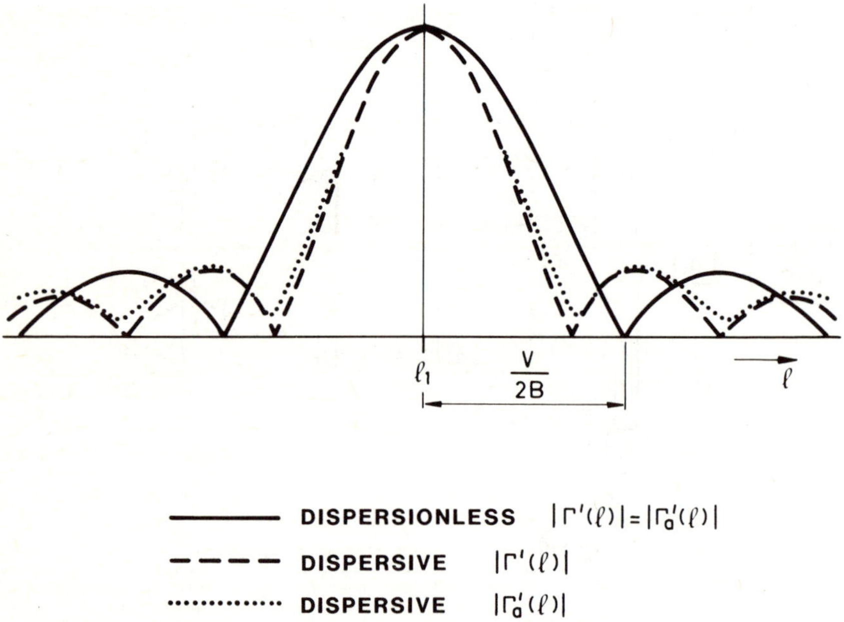

Fig. 6.11. *Comparison of computed synthetic pulse time-domain responses of a locating reflectometer operating in*
——— a dispersionless line
– – – – waveguide with wavenumber swept linearly
······· waveguide with frequency swept linearly

The reference reflection Γ_l in the second arm of the reflectometer can be a sliding short circuit with a reciprocating motion which controls the l axis of a display of the locating function. An alternative arrangement in waveguide is to form a 270° E-plane circular bend with a continuous longitudinal slot cut in the centre of the inner broad wall. A reflecting probe can be inserted through the slot and moved along the waveguide using a motor with its axis at the centre of the circle [23], and Somlo also introduced a resonant iris into the circuit to weight the frequency response of the reflectometer so as to suppress the sidelobes of the sinc function response. With this frequency weighting and a linear frequency sweep, the response of the locating reflectometer is denoted by $\Gamma'_w(l)$.

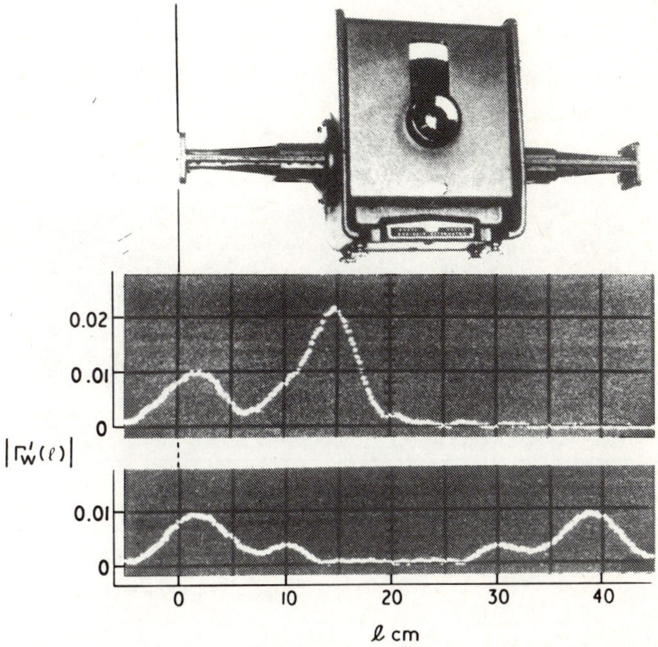

Fig. 6.12. *Internal reflections of a rotary-vane attenuator as displayed by the locating reflectometer of Somlo* [23].
The upper trace corresponds to maximum attenuation, and the lower trace to minimum attenuation

[Reproduced by permission of IEEE]

Some results obtained with the locating reflectometer in X-band waveguide are shown in Figs. 6.12–6.14. The internal reflections of a rotary vane attenuator are shown in magnitude and location in Fig. 6.12, for both the maximum and minimum attenuation settings. Fig. 6.13 shows the reflection arising as a result of dimensional differences and losses at a flange [24]. The internal reflections of a horn antenna are shown in Fig. 6.14. Because the locating reflectometer displays the reflections as a function of distance, it was not necessary to position the antenna in a non-reflecting environment to obtain the curves. The upper trace in Fig. 6.14 shows the reflections before the introduction of steel balls into the antenna. These balls were moved about using external magnets, and then glued in positions which reduced $|\Gamma'_w(l)|$ to that shown in the lower trace in Fig. 6.14, thereby providing a broadband match.

The circuit in Fig. 6.10, with the omission of the integrators, could be used to determine the position of a discontinuity if white noise of sufficient power were used as the source transmission. The operation is analogous to a white light interferometer in which all spectral components of a white light source reflected from the discontinuity interfere constructively with a reference beam traversing an equal path length [25, 26].

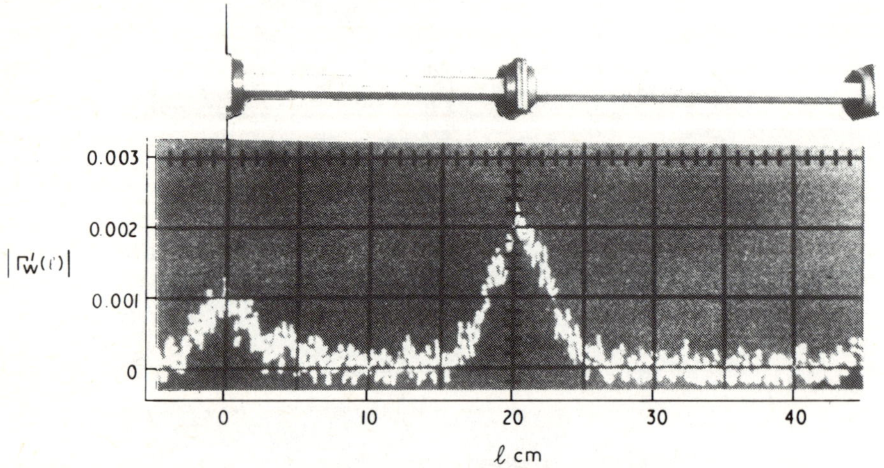

Fig. 6.13. *Locating reflectometer display of the reflections set up at waveguide flanges*
[Reproduced by permission of IEEE]

The locating reflectometer can also be used to synthesize the reflected or transmitted pulse response of a dispersive or dispersionless system. If the second arm of the reflectometer is a dispersionless line, signal B is proportional to $\Gamma_l \exp(-\mathrm{j}4\pi fl/v)$. For a frequency sweep which is linear with time and between the limits f_1 and f_2, the output of one integrator is then proportional to $f_r(2l/v)$ since

$$f_r\left(\frac{2l}{v}\right) = 2\,\mathrm{Re}\int_{f_1}^{f_2} \Gamma(f)\,\mathrm{e}^{\mathrm{j}\,4\pi fl/v}\,\mathrm{d}f \tag{6.44}$$

The second arm must be dispersionless in order to synthetically construct the distorted response of a dispersive system which would have been observed if the system had been tested using a band-limited impulse, as discussed in Section 6.1.6.

If the DUT is a two-port, and signal A is taken from the output port, the transmitted pulse response of the device is synthesized in the same manner. An example of such a measurement was given by Somlo [27] for a transmission type cavity resonator, and the synthesized transmission is shown in Fig. 6.15. The synthesized exciting pulse corresponding to the swept bandwidth is shown as the dotted curve, and its width at half-height would have been less than 300 ps if transmitted in real-time. Of course, the curve in Fig. 6.15 was recorded much more slowly using an X–Y recorder with the X displacement proportional to l, and the Y input connected to the (real) integrator output. The Q-factor computed from the logarithmic decrement of the response curve agreed well with that found using a half-power bandwidth method.

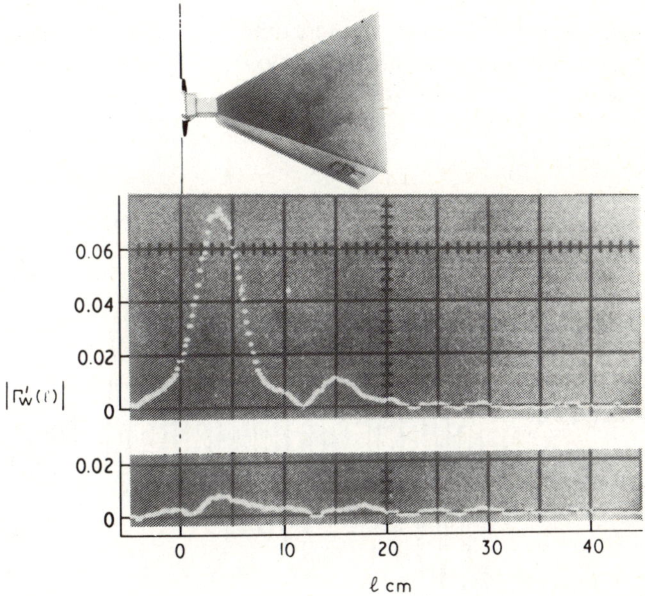

Fig. 6.14. ***An example of the use of the locating reflectometer to measure the internal reflections of antennas. Because the axial resolution was approximately 4 cm, room reflections did not interfere with the measurement***

The upper trace shows the original reflections in an uncompensated horn, and after introducing compensating elements while monitoring the display, the reflections were reduced to those of the lower trace

[Reproduced by permission of IEEE]

In summary, the locating reflectometer produces the locating function when the signal is swept linearly in wavenumber, and the reference transmission line has the same dispersive characteristic as the DUT. When the DUT is dispersive and the signal is swept linearly in frequency, resolving ability is slightly degraded. To produce a synthesized pulse response from dispersive or dispersionless devices, the second arm of the reflectometer must be a dispersionless transmission line, and the signal must be swept linearly in frequency.

6.2.3 Deconvolution to improve axial resolution

It is clear from eqn. 6.25 that if the locating function is evaluated using data derived over a finite bandwidth, a lumped discontinuity with reflection coefficient

Γ_{l_1} will produce a distributed locating display. This 'smearing' of the true axial distribution $\Gamma_{l_1}\ \delta(l-l_1)$ causes the locating function for several small localized discontinuities to be the superposition of several distributed functions, assuming interaction between the discontinuities is not significant. The 'smearing' process is the result of the convolution of the true axial distribution $h(l)$ with the so-called instrument function (pulse response, aperture function) $g(l)$, and is given by

$$\Gamma'(l) = h(l) * g(l) \tag{6.45}$$

Only if the instrument function is an impulse function (requiring infinite bandwidth) is the locating function the same as the true axial distribution. The aim of deconvolution is to recover $h(l)$ from $\Gamma'(l)$.

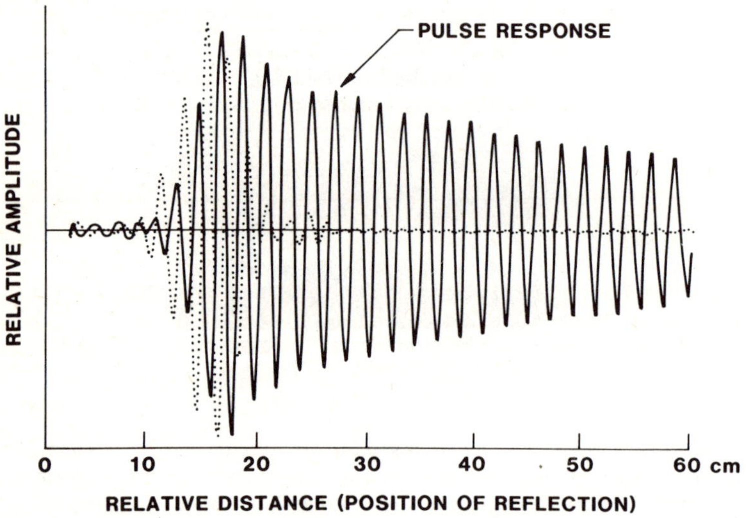

Fig. 6.15. *Use of the locating reflectometer to obtain the pulse response of a transmission type cavity. Short-term phenomena such as the build-up and exponential decay rates of the high Q cavity were recorded using a low-speed chart recorder*

[Reproduced by permission of IEEE]

The convolution theorem states that to convolve two functions is to multiply their Fourier transforms. Therefore, if the Fourier transform of the locating function is divided by the Fourier transform of the instrument function, the result is the Fourier transform of the true axial distribution, from which $h(l)$ can be recovered.

This deconvolution process was applied in X-band waveguide by Somlo [28] using the locating reflectometer. The instrument function was taken as the locating function for a small conductive ball (bearing ball) centrally located on the broad wall of terminated waveguide. Such a discontinuity approximates an obstacle of zero width, and is known to produce a nearly constant reflection across the waveguide band [29]. The Fourier transform of the instrument function was calculated, as were the transforms of the real and imaginary parts of the locating function obtained from the locating reflectometer for the various waveguide devices of interest. In all cases, spectral components outside the swept frequency range were set to zero.

The locating function is analytic, with real and imaginary parts which are a Hilbert transform pair, and with a spectrum which exists only for positive frequencies. Consequently the complex function $h(l)$ can be computed after dividing the transform of the real part of $\Gamma'(l)$ by the transform of $g(l)$ and discarding negative frequency spectral components. Alternatively, $h(l)$ can be determined from the imaginary part of $\Gamma'(l)$. If both these results for $h(l)$ are averaged, system noise effects are reduced.

Figs. 6.16 to 6.18 show the resolution improvement obtained by deconvolution. Fig. 6.16 shows the locating function for two small discontinuities, which before deconvolution appear to be 3·8 cm apart. After deconvolution the true separation of 3 cm is revealed. Fig. 6.17 shows the locating function for a waveguide twist. After deconvolution the locations of individual discontinuities are found to be in the planes where the twist starts and ends. Fig. 6.18 shows how individual coupling apertures of a cross-guide coupler can be revealed after deconvolution, in this case with an instrument function obtained from a single coupling aperture.

It is evident, particularly from Fig. 6.16, that deconvolution does not fully recover the true axial distribution of discontinuities. Apart from the inescapable limitation of finite bandwidth, the assumption of superposition is only approximate, because of interaction between discontinuities. Even if the discontinuities are small so that multiple reflections are low, close spacing causes the evanescent fields surrounding the discontinuities to interact.

6.2.4 *Locating function from a Smith chart display*

A semigraphical method for finding $\Gamma'(l)$ from a Smith chart display of $G(\nu)$ has been described by Somlo [30], and is based on approximating the integral in eqn. 6.18 by the summation

$$\Gamma'(l) \simeq \frac{1}{N}\sum_{n=1}^{N} G(\nu_n)\,\mathrm{e}^{\mathrm{j}4\pi\nu_n l} \qquad (6.46)$$

where the reflection coefficient is known at N values ν_n equally spaced by $\Delta\nu$ between ν_1 and ν_N.

$G(\nu_n)$ are the reflection coefficient vectors for the DUT as measured in a plane designated $l = 0$, and can be plotted in magnitude and phase on a Smith chart. It follows from eqn. 6.46 that the mean of these vectors approximates $\Gamma'(0)$. Each vector $G(\nu_n)$ can be multiplied graphically by $\exp(\mathrm{j}4\pi\nu_n l_1)$ by rotating it anti-clockwise through an angle $\theta_n = 4\pi\nu_n l_1$ on the Smith chart, and the mean of the rotated vectors is, from eqn. 6.46, an approximation to $\Gamma'(l)$, as shown in Fig. 6.19. The calculation of the rotation angles and the mean are simple operations which

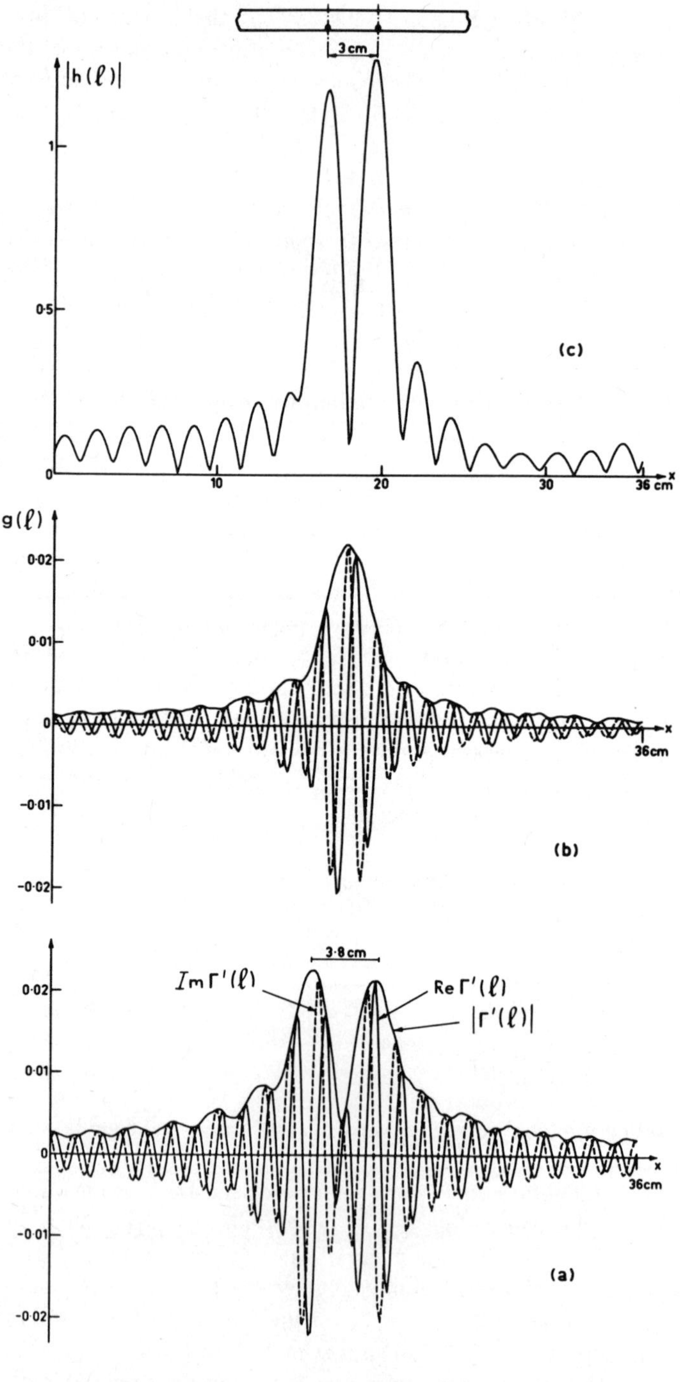

Fig. 6.16. *Improvement in resolving ability of the locating reflectometer using deconvolution*
The two small discontinuities 3 cm apart were accurately located (*c*) after deconvolving the observed function (*a*) with the instrument function (*b*)
[Reproduced by permission of IEEE]

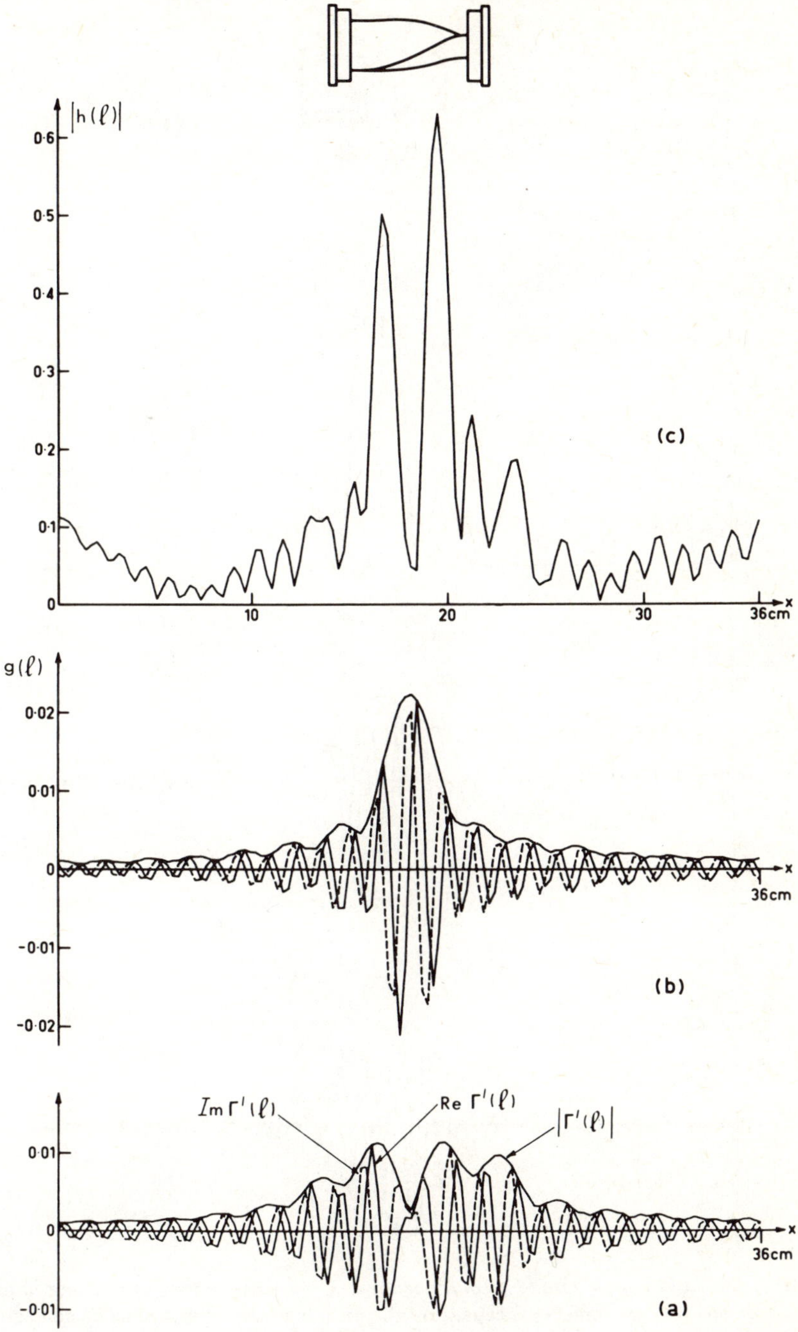

Fig. 6.17. *Improvement in resolving ability of the locating reflectometer using deconvolution*
The internal reflections of a short waveguide twist were resolved and accurately located (*c*) after deconvolving the observed function (*a*) with the instrument function (*b*)

[Reproduced by permission of IEEE]

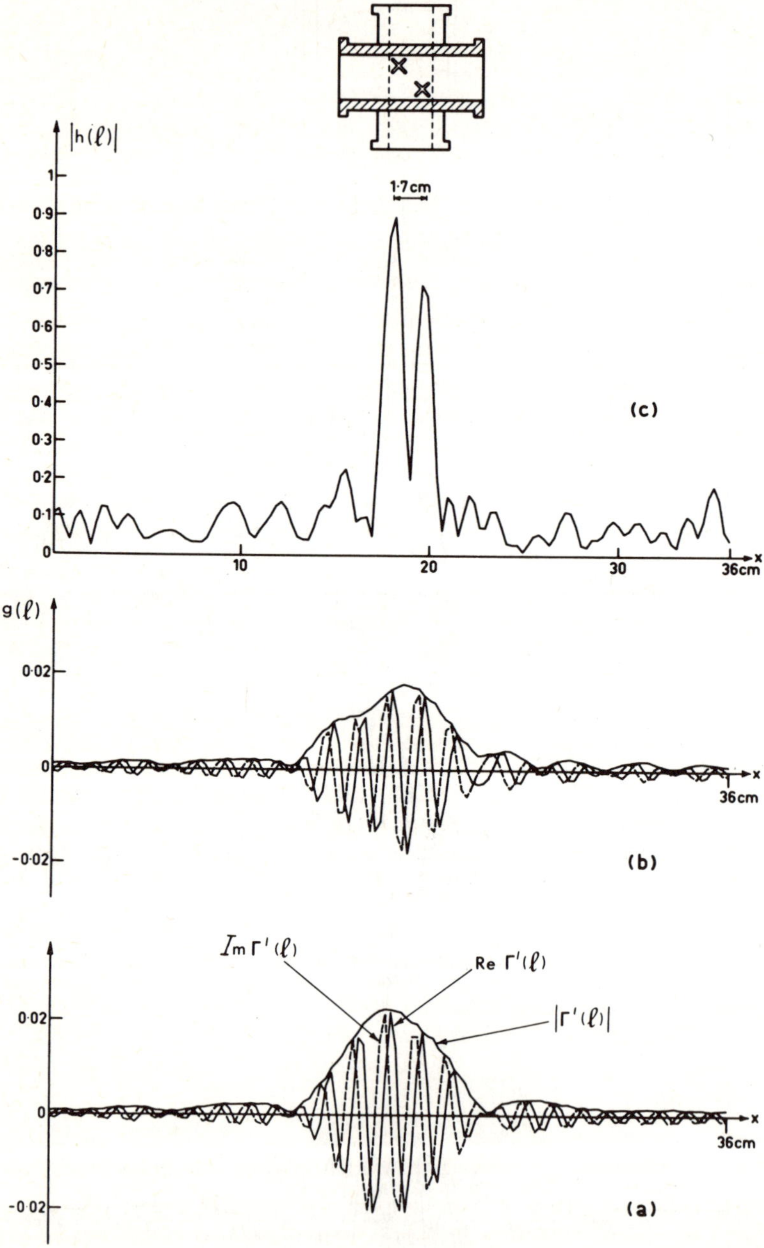

Fig. 6.18. *Improvement in resolving ability of the locating reflectometer using deconvolution*
The internal reflections caused by the coupling holes in a cross-guide coupler were resolved and accurately located (*c*) after deconvolving the observed function (*a*) with the instrument function (*b*)

[Reproduced by permission of IEEE]

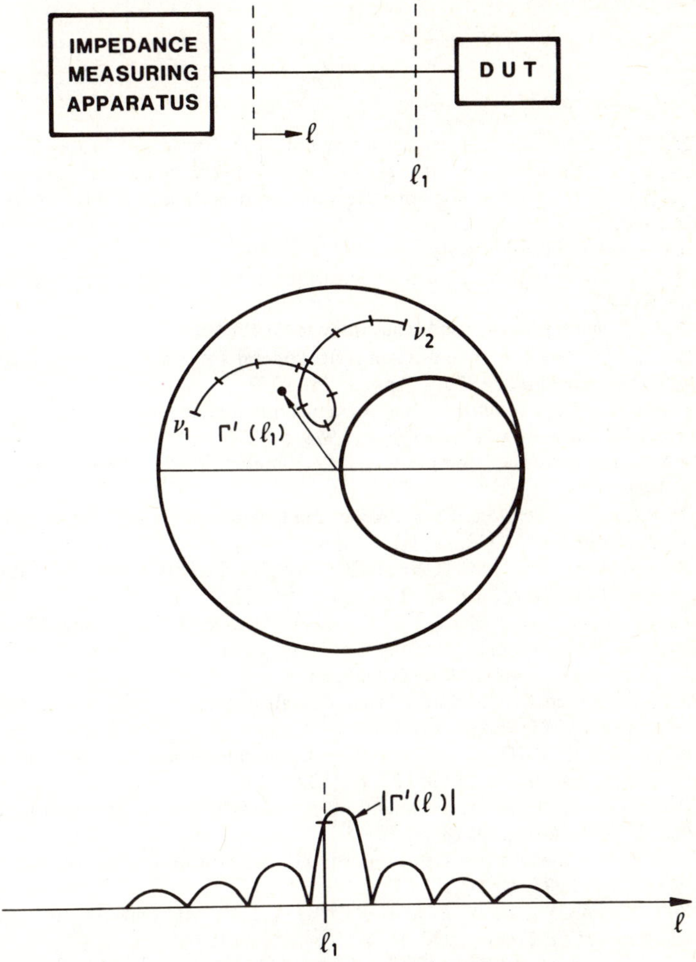

Fig. 6.19. *Locating reflectometry from Smith chart plotting*
$\Gamma'(\ell)$ is the mean of $\Gamma(\nu)$ plotted on the Smith chart referenced to the location ℓ_1

can be done with ease on a pocket calculator. The maximum value of $|\Gamma'(l)|$ occurs for that value of l which rotates the $G(\nu_n)$ to be most nearly coincident.

This procedure enables synthetic TDR measurements to be performed with very little equipment – perhaps only a slotted line. However, it must be remembered that because $G(\nu)$ is known at only N points, the expression in eqn. 6.46 will be periodic in l, with period

$$d = \frac{1}{2\Delta\nu} \tag{6.47}$$

If the reflection coefficients are measured at equal intervals in frequency, the above procedure approximates $\Gamma'_{\mathbf{a}}(l)$.

References

1 BRACEWELL, R. N.: 'The Fourier transform and its applications' (McGraw-Hill 1965)
2 OLIVER, B. M.: 'Time-domain reflectometry', *Hewlett-Packard Journal*, 1964, **15**
3 STRICKLAND, J. A.: 'Time-domain reflectometry measurements' (Tektronix, Measurement Concepts)
4 Hewlett-Packard Application Note 62, 1964
5 PILLER, U.: 'Time domain immittance measurements', 4th European Microwave Conf., 1974, pp. 61–63
6 COWAN, G.: 'Digital processing adds accuracy to TDR', *Microwaves*, 1975, **14**, pp. 47–51
7 ELLIOTT, B. J.: 'Sensitive time domain reflectometer for observing multiple reflections', *IBM Tech. Disclosure Bull.*, 1975, **17**, pp. 3078–3079
8 ELLIOTT, B. J.: 'High resolution time domain reflectometer for measuring attofarad and femtohenry discontinuities', *CPEM Digest*, 1976, p. 22
9 HOLLWAY, D. L.: 'The Comparison reflectometer', *IEEE Trans.*, 1967, **MTT-15**, pp. 250–259
10 CHAMBERS, B.: 'Application of super-resolution techniques to time domain reflectometry systems', *Electron. Lett.*, 1978, **14**, pp. 97–98
11 MACKAY, N. A. M. and PENSTONE, S. R.: 'A high-sensitivity narrow-band time-domain reflectometer', *IEEE Trans.*, 1974, **IM-23**, pp. 155–158
12 NICOLSON, A. M., BENNETT, Jr. C. L., LAMENSDORF, D. and SUSMAN, L.: 'Applications of time-domain metrology to the automation of broad-band microwave measurements', *IEEE Trans.*, 1972, **MTT-20**, pp. 3–9
13 ANDREWS, J. R. and GANS, W. L.: 'Time domain automatic network analyser', *Mesure, L'onde electrique*, 1975, **55**, pp. 569–574
14 RIGG, P. R. and CARROLL, J. E.: 'Low-cost computer-based time-domain microwave network analyser', *IEE Proc.*, 1980, **127**, pp. 107–111
15 CORMACK, G. D.: 'TDR measurement of thickness of conductive plating on wires', *IEEE Trans.*, 1974, **IM-23**, pp. 181–182
16 KENT, M.: 'Time domain measurements of the dielectric properties of frozen fish', *J. Microwave Power*, 1975, **10**, pp. 37–48
17 ROBINSON, L. A., WEIR, W. B. and YOUNG, L.: 'An RF time-domain reflectometer not in real time', *IEEE Trans.*, 1972, **MTT-20**, pp. 855–857
18 HINES, M. E. and STINEHELFER Sr, H. E.: 'Time-domain oscillographic microwave network analysis using frequency domain data', *IEEE Trans.*, 1974, **MTT-22**, pp. 276–282
19 HANSEN, W. W.: *in* 'Radar system engineering' RIDENOUR, L. N. (ed.) MIT Ser. No. 1, (McGraw-Hill, 1947), pp. 143–150
20 YAMAURA, I. and HIDAKA, T.: 'The double-swept-frequency locating reflectometer', *IEEE Trans.*, 1975, **MTT-23**, pp. 316–317
21 SOMLO, P. I. and HOLLWAY, D. L.: 'Microwave locating reflectometer', *Electron. Lett.*, 1969, **5**, pp. 468–469
22 DETLEFSEN Von, J.: 'Reflexionsstellenortung an Messobjekten mit Leitungscharakter durch Fourier Transformation des Reflexionfaktors', *NTZ*, 1972, **6**, pp. 269–274
23 SOMLO, P. I.: 'The locating reflectometer', *IEEE Trans.*, 1972, **MTT-20**, pp. 105–112
24 SOMLO, P. I.: 'The effect of flange loss on the reflection coefficient of reduced-height waveguide reflection standards', *IEEE Trans.*, 1979, **MTT-27**, pp. 795–797
25 GEBBIE, H. A., OSBORNE, C. F., STONE, N. W. B. and TAYLOR, K. B.: 'The N. P. L. Teramet', *The Engineer*, 1968

26 GEBBIE, H. A., OSBORNE, C. F. and STONE, N. W. B.: 'Radar for engineers', *New Scientist*, 1968, **40**

27 SOMLO, P. I.: 'Display of microwave pulse response via the real-time Fourier transform of the transfer function', *Electron. Lett.*, 1972, **8**, p. 631

28 SOMLO, P. I.: 'Band-limited deconvolution of Locating reflectometer results', *IEEE Trans.*, 1976, **MTT-27**, pp. 128–135

29 SOMLO, P. I. and HOLLWAY, D. L.: 'Conductive contacting spheres on the centre of the broad wall of rectangular waveguides', *Electron. Lett.*, 1972, **8**, pp. 507–508

30 SOMLO, P. I.: 'Time-domain reflectometry from Smith-chart plotting', *Electron. Lett.*, 1973, **9**, pp. 221–222

Chapter 7

Outline of automated microwave impedance measuring methods

As mentioned in the preface, a detailed description of automated microwave measurement systems is outside the scope of this book, although many of the techniques discussed in earlier chapters are amenable to automation. By common usage, an automated microwave measuring apparatus has come to be understood as a system which is calibrated by external standards, and which automatically applies the results of those calibrations to the quantity being measured, before presenting that quantity to the operator. Such a calibration procedure treats the apparatus as a single device, and models it as such.

The availability of microprocessors and dedicated computers has brought about a change in the philosophy of microwave measurements. Whereas in the past much effort was expended on building more nearly perfect components in order to perform more accurate measurements, it is now practical to use less than perfect components in a system whose overall characteristics are determined by calibration of the system as a whole. Thus the emphasis has shifted from trying to produce a nearly perfect measuring apparatus to producing well-defined calibrating standards – a much easier task.

The two types of equipment currently classed as automated impedance measurement apparatus are the automatic network analysers and six-port systems.

7.1 Automatic network analysers

Automatic network analysers (ANA) are available in two types: scalar automatic network analysers (SANA) and vector automatic network analysers (VANA). Because a SANA measures only the magnitude of the required quantity, it can be built with scalar detectors. Inexpensive diode detectors may be used, but they limit the dynamic range of the instrument [1].

A schematic diagram of the more versatile (and expensive) VANA is shown in Fig. 7.1. With additional switching, multiport networks can be analysed for both reflection and transmission characteristics, as shown in Fig. 7.2. A sample of the input signal is taken as a reference, and, along with a sample of the reflected (or transmitted) signal, is frequency converted to an intermediate frequency (IF), either by coherent sampling or phase-locked superheterodyne mixing. The two IF signals are measured using a vectorial ratio meter. After digitization, the corrections necessary to account for the characteristics of the system are applied, and the measured (and corrected) quantity can be displayed on a CRO or plotter, printed in digital form, or stored for further digital processing. Frequency conversion is unavoidable with a VANA since no phase meter can operate at all the microwave frequencies likely to be required. Frequency conversion also has the advantage of improving the dynamic range.

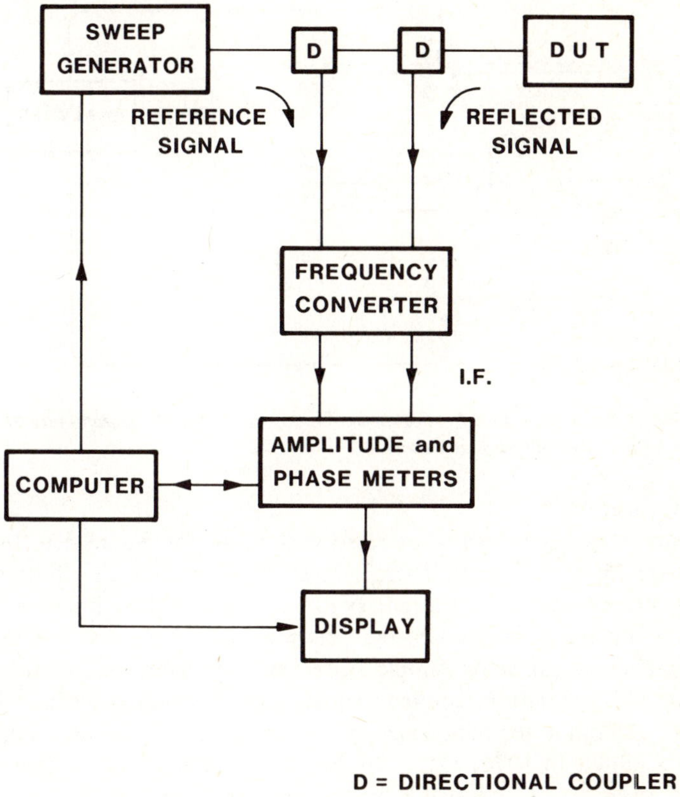

Fig. 7.1. *Vector automatic network analyser system*

Modern dedicated microprocessors make very high-speed ANA operation possible. For example, one commercially available VANA is capable of measuring the *S*-parameters of a two-port network at 400 frequencies in about one second.* In that time, the results are computer corrected using the information obtained

* Hewlett Packard HP 8510 series.

from prior calibration of the analyser, and the Fourier transform is evaluated, making the results available in both the frequency and time-domains.

The errors of an ANA can be classified as either microwave or electronic in origin.

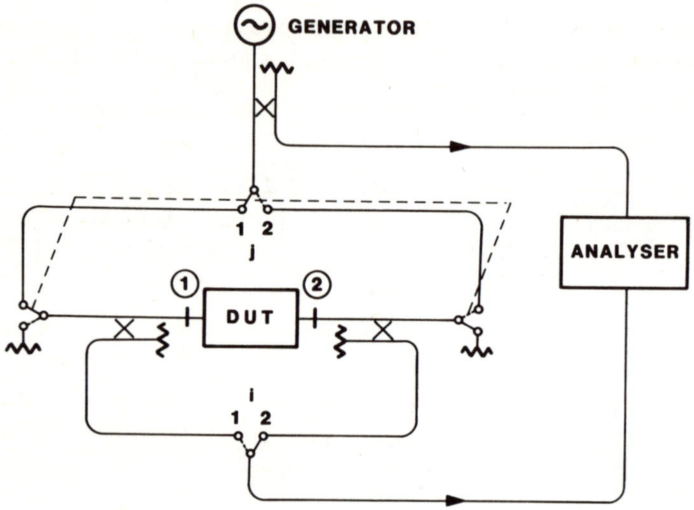

Fig. 7.2. *Switching arrangement for the measurement of the four S_{ij} parameters of a two-port, using a vector network analyser*

(*a*) Microwave errors

The measured quantity must be corrected to account for the imperfections of the microwave circuit components. These imperfections include the finite directivity of the couplers, mismatched ports, and switch leakages. Although the quality of the microwave components is unimportant in principle, in practice the use of high quality components can make simple system models viable, and therefore reduce the amount of computation required. Numerous calibration procedures have been proposed to determine the model elements. A comprehensive bibliography of ANA calibration methods to 1976 is given by Beatty [2], and a cross-section of various subsequent developments is available in the literature [3–9].

(*b*) Electronic errors

Apart from the microwave errors, electronic errors are introduced by the complicated circuitry associated with a VANA, and to a lesser extent a SANA. In a VANA, the frequency convertor, voltmeter, and phasemeter units have their own errors which may be non-linear, and therefore not easily accounted for by the calibration procedure. These units set a minimum uncertainty for the whole ANA, and the best results claimed for commercially available systems are about $\pm 0{\cdot}006$ in $|\Gamma|$, and about ± 1 degree in $\arg(\Gamma)$ up to 18 GHz.

7.2 Six-port systems

An alternative type of automated measurement apparatus can be constructed using a microwave network with six ports, and applying the theory developed by Engen and Hoer [10–14]. If port 1 of the linear six-port network in Fig. 7.3 is driven by a monochromatic signal, the reflection coefficient of a termination connected to port 2 can be determined in both magnitude and phase from the powers sensed at the other four ports. The advantages of this type of reflectometer include:

(*a*) the absence of any need for frequency conversion and phasemeters;
(*b*) the absence of any need to match power meters or calibrate them to read power in an absolute sense. They need only be linear and present a constant impedance;
(*c*) for all practical purposes, the absence of electronic errors;
(*d*) the ability to measure transmission characteristics, by introducing a second six-port. Such networks are called 'dual six-ports', and various calibration procedures have been reported [15–18].

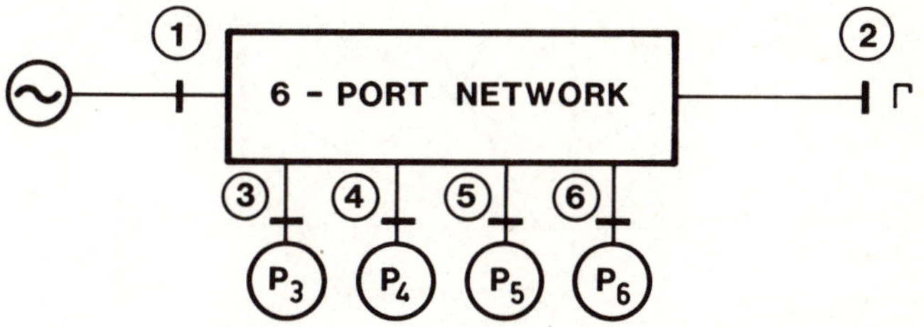

Fig. 7.3. *Six-port reflectometer system. The unknown reflection coefficient Γ present at port 2 is uniquely determined by the four powers measured at ports 3–6*

The disadvantages of six-ports include:

(*a*) the need to filter spurious signals and harmonics because of the inherently wideband response of the power meters,
(*b*) the limited dynamic range of the power meters. Although this is not a problem for most reflectometry work using a well designed six-port junction, it can cause difficulties in transmission measurements through lossy networks.*
(*c*) If thermistor power meters are used to achieve good linearity, high-power sources are needed. This has resulted in some criticism of six-ports [20–22]. However, the use of semiconductor diodes as power meters eliminates this disadvantage [19, 23], and with diode linearization, effective non-linearities of less

* Contrary to some opinions expressed in the literature, the accurate measurement of small reflection coefficients is not limited by power meter dynamic range, but rather by resolution. Six-ports can be designed so that for all passive reflections the dynamic range is below (say) 16 dB [19].

than 0·001 dB are obtainable [19]. Alternatively, unlinearized diodes may be used as level sensings elements without accuracy degradation [24].

7.2.1 Six-port reflectometer theory

A linear six port junction can be represented by the signal flow graph in Fig. 7.4. Each of the 36 signal paths is described by a member of the S-parameter set associated with characteristic port impedances ξ_i, $i = 1, \ldots, 6$.*

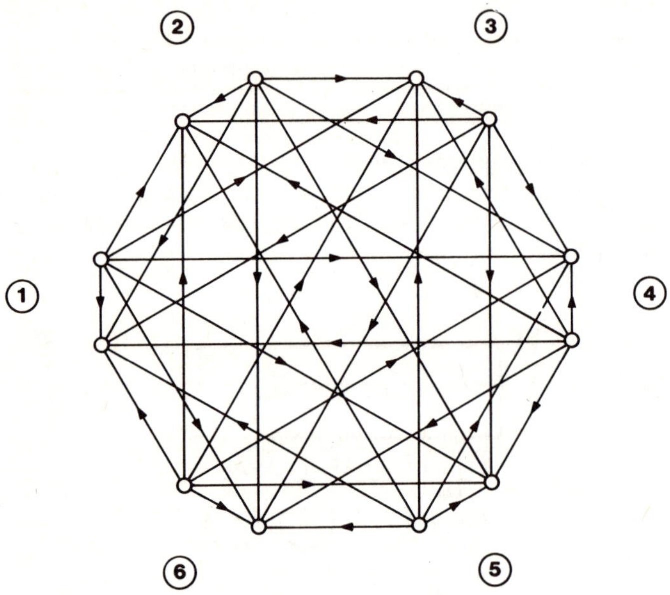

Fig. 7.4. *General flow graph representation of a six-port network*

When the six-port junction is used as a reflectometer, a generator with reflection coefficient Γ_G is connected to port 1, the unknown with reflection coefficient Γ is connected to port 2, and four power meters with impedances z_i are connected to ports i, $i = 3, 4, 5, 6$. We are at liberty to choose $\xi_i = z_i$, and then, provided the (unknown) z_i remain constant with variation of incident power, the signal flow graph reduces to that in Fig. 7.5.

The signal b_i emergent at port i is found by applying the flow graph reduction techniques of Section 2.3.1, and is given by

$$b_i = \frac{S_{i1} + \Gamma(S_{i2}S_{21} - S_{22}S_{i1})}{(1 - \Gamma_G S_{11})(1 - \Gamma S_{22} - \Gamma\Gamma_G S_{21}S_{12})}, \quad i = 2, \ldots, 6 \qquad (7.1)$$

* As discussed in Section 2.1.1, S parameters are defined only in conjunction with characteristic port impedances.

and in particular

$$b_2 = \frac{S_{21}}{(1-\Gamma_G S_{11})(1-\Gamma S_{22} - \Gamma\Gamma_G S_{21} S_{12})} \tag{7.2}$$

Since

$$\Gamma = \frac{a_2}{b_2} \tag{7.3}$$

b_i may be expressed in terms of the incident and emergent signals at port 2 as

$$b_i = M_i a_2 + N_i b_2 \tag{7.4}$$

where the complex constants M_i, N_i are

$$M_i = S_{i2} - \frac{S_{i1} S_{22}}{S_{21}}$$
$$N_i = \frac{S_{i1}}{S_{21}} \tag{7.5}$$

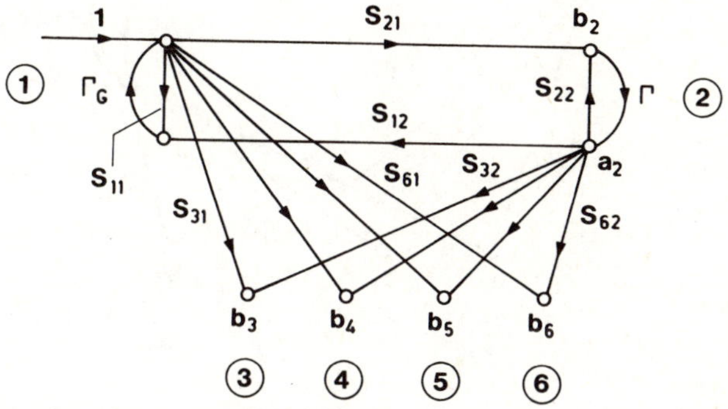

Fig. 7.5. *Flow graph representation of a six-port network with the characteristic impedances of ports 3–6 designated equal to those of the power meters connected them. A generator with reflection coefficient Γ_G is connected to port 1, and a termination with reflection coefficient Γ is connected to port 2*

Notice that eqn. 7.4 is independent of Γ_G. This occurs because port 1 is simply a driving port for the network, and the fact that the signal incident on port 1 is a composite of generator output and multiple reflections due to generator mismatch, is of no consequence in determining the relationships between signals inside the network. The simplified flow graph in Fig. 7.6 is therefore sufficient to describe the operation of the six-port, provided it is used as a reflectometer. (Of course, if the generator impedance is known to be constant, then ξ_1 may be chosen equal to it, and Fig. 7.6 is a complete description of the network.)

The power indicated by the power meter connected to port i is proportional to the square of the magnitude of the signal incident upon it. Therefore the ratio p_i of power indicated at port i to that indicated at a reference port (say port 6) is

$$p_i = K_i \frac{|b_i|^2}{|b_6|^2} = K_i \left| \frac{N_i + \Gamma M_i}{N_6 + \Gamma M_6} \right|^2, \quad i = 3, 4, 5 \tag{7.6}$$

where the K_i are real constants. Provided $N_i \neq 0$, this expression can be simplified to

$$p_i = q_i \left| \frac{1 + A_i \Gamma}{1 + A_6 \Gamma} \right|^2, \quad i = 3, 4, 5 \tag{7.7}$$

where

$$\left. \begin{aligned} q_i &= K_i \frac{|S_{i1}|^2}{|S_{61}|^2} \\ A_i &= \frac{S_{i2} S_{21}}{S_{i1}} - S_{22} \end{aligned} \right\} \tag{7.8}$$

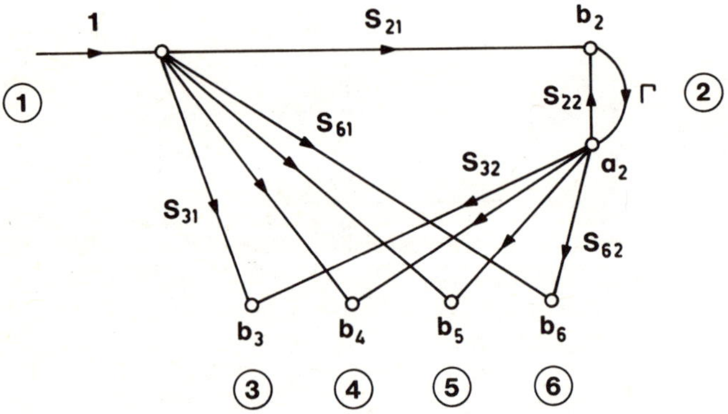

Fig. 7.6. *Flow graph sufficient to describe the operation of a six-port reflectometer. Because the reflectometer analysis is independent of the reflection coefficient of the generator, some paths shown in Fig. 7.5 can be omitted.*

The three expressions in eqns. 7.7 describe relationships between the power meter ratios and the unknown reflection coefficient Γ in terms of three real constants q_i, and four complex constants A_i. Let us assume that a calibration procedure is available which allows us to determine these constants. We shall now seek a solution to eqns. 7.7 with Γ as the unknown. Such a solution will enable us to compute Γ from the constants q_i, A_i, and the power meter ratios p_i measured when the unknown termination is connected to the measuring port, port 2.

First, write Γ and A in terms of their real and imaginary components as

$$\Gamma = x + jy; \quad A_i = \alpha_i + j\beta_i \tag{7.9}$$

Then eqns. 7.7 may be expanded as

$$|\Gamma|^2(p_i|A_6|^2 - q_i|A_i|^2) + 2x(p_i\alpha_6 - q_i\alpha_i) - 2y(p_i\beta_6 - q_i\beta_i)$$
$$= q_i - p_i, \quad i = 3, 4, 5 \tag{7.10}$$

Since

$$|\Gamma|^2 = x^2 + y^2 \tag{7.11}$$

eqns. 7.10 give three non-linear equations in x and y. These may be solved for x and y by treating $|\Gamma|^2$, x, y as three unknowns, and hence Γ can be written in terms of eleven real constants as

$$\Gamma = \frac{\sum_{i=3}^{6}(F_i + jG_i)p_i}{1 + \sum_{i=3}^{5} H_i p_i} \tag{7.12}$$

where

$$F_3 = \frac{-1}{2Dq_3}[|A_4|^2(\beta_5 - \beta_6) + |A_5|^2(\beta_6 - \beta_4) + |A_6|^2(\beta_4 - \beta_5)]$$

$$F_4 = \frac{1}{2Dq_4}[|A_5|^2(\beta_6 - \beta_3) + |A_6|^2(\beta_3 - \beta_5) + |A_3|^2(\beta_5 - \beta_6)]$$

$$F_5 = \frac{-1}{2Dq_5}[|A_6|^2(\beta_3 - \beta_4) + |A_3|^2(\beta_4 - \beta_6) + |A_4|^2(\beta_6 - \beta_3)]$$

$$F_6 = \frac{1}{2D}[|A_3|^2(\beta_4 - \beta_5) + |A_4|^2(\beta_5 - \beta_3) + |A_5|^2(\beta_3 - \beta_4)]$$

$$G_3 = \frac{-1}{2Dq_3}[|A_4|^2(\alpha_5 - \alpha_6) + |A_5|^2(\alpha_6 - \alpha_4) + |A_6|^2(\alpha_4 - \alpha_5)]$$

$$G_4 = \frac{1}{2Dq_4}[|A_5|^2(\alpha_6 - \alpha_3) + |A_6|^2(\alpha_3 - \alpha_5) + |A_3|^2(\alpha_5 - \alpha_6)]$$

$$G_5 = \frac{-1}{2Dq_5}[|A_6|^2(\alpha_3 - \alpha_4) + |A_3|^2(\alpha_4 - \alpha_6) + |A_4|^2(\alpha_6 - \alpha_3)]$$

$$G_6 = \frac{1}{2D}[|A_3|^2(\alpha_4 - \alpha_5) + |A_4|^2(\alpha_5 - \alpha_3) + |A_5|^2(\alpha_3 - \alpha_4)]$$

$$H_3 = \frac{-1}{Dq_3}[|A_4|^2(\alpha_5\beta_6 - \alpha_6\beta_5) + |A_5|^2(\alpha_6\beta_4 - \alpha_4\beta_6)$$
$$+ |A_6|^2(\alpha_4\beta_5 - \alpha_5\beta_4)]$$

$$H_4 = \frac{1}{Dq_4}[|A_5|^2(\alpha_6\beta_3 - \alpha_3\beta_6) + |A_6|^2(\alpha_3\beta_5 - \alpha_5\beta_3)$$
$$+ |A_3|^2(\alpha_5\beta_6 - \alpha_6\beta_5)]$$

$$H_5 = \frac{-1}{Dq_5}[|A_6|^2(\alpha_3\beta_4 - \alpha_4\beta_3) + |A_3|^2(\alpha_4\beta_6 - \alpha_6\beta_4)$$
$$+ |A_4|^2(\alpha_6\beta_3 - \alpha_3\beta_6)]$$

$$D = |A_3|^2(\alpha_4\beta_5 - \alpha_5\beta_4) + |A_4|^2(\alpha_5\beta_3 - \alpha_3\beta_5) + |A_5|^2(\alpha_3\beta_4 - \alpha_4\beta_3)] \quad (7.13)$$

That the unknown reflection coefficient can be found as a function of power ratios is perhaps not surprising, since reflection coefficients are themselves ratios. It is also convenient, because the power meters are required only to present a constant impedance to the six-port, and respond linearly with input power. They need not be calibrated to read power in an absolute sense.

7.2.1.1. Alternative formulation: The derivation of eqns. 7.7 from eqns. 7.6 is valid only if $N_i \neq 0$, $i = 3, 4, 5, 6$. Consequently, eqns. 7.7 are unsuitable if any of the reflectometer power meters responds solely to the reflected signal a_2 ($S_{i1} = 0$). In such a case, an alternative simplification of eqns. 7.6 gives

$$p_i = r_i \left| \frac{\Gamma - B_i}{\Gamma - B_6} \right|^2, \quad i = 3, 4, 5 \quad (7.14)$$

where

$$r_i = K_i \left| \frac{S_{i1}S_{22} - S_{i2}S_{21}}{S_{61}S_{22} - S_{62}S_{21}} \right|^2 \quad (7.15)$$

and

$$B_i = \frac{S_{i1}}{S_{i1}S_{22} - S_{i2}S_{21}} \quad (7.16)$$

and is valid provided $M_i \neq 0$, $i = 3, 4, 5, 6$. This condition will be violated if any of the power meters responds solely to the forward signal b_2 at the measuring port.

We now seek a solution to eqns. 7.14. These three equations can be written as

$$|\Gamma|^2(p_i - r_i) - 2x(p_i\gamma_6 - r_i\gamma_i) - 2y(p_i\delta_6 - r_i\delta_i) = r_i|B_i|^2 - p_i|B_6|^2, \quad i = 3, 4, 5 \quad (7.17)$$

where

$$B_i = \gamma_i + j\delta_i \quad (7.18)$$

After solving for x and y, Γ is given by eqn. 7.12 with

$$F_3 = \frac{-1}{2Er_3}[|B_4|^2(\delta_5 - \delta_6) + |B_5|^2(\delta_6 - \delta_4) + |B_6|^2(\delta_4 - \delta_5)]$$

$$F_4 = \frac{1}{2Er_4}[|B_5|^2(\delta_6 - \delta_3) + |B_6|^2(\delta_3 - \delta_5) + |B_3|^2(\delta_5 - \delta_6)]$$

$$F_5 = \frac{-1}{2Er_5}[|B_6|^2(\delta_3 - \delta_4) + |B_3|^2(\delta_4 - \delta_6) + |B_4|^2(\delta_6 - \delta_3)]$$

$$F_6 = \frac{1}{2E}[|B_3|^2(\delta_4 - \delta_5) + |B_4|^2(\delta_5 - \delta_3) + |B_5|^2(\delta_3 - \delta_4)]$$

$$G_3 = \frac{1}{2Er_3}[|B_4|^2(\gamma_5-\gamma_6)+|B_5|^2(\gamma_6-\gamma_4)+|B_6|^2(\gamma_4-\gamma_5)]$$

$$G_4 = \frac{-1}{2Er_4}[|B_5|^2(\gamma_6-\gamma_3)+|B_6|^2(\gamma_3-\gamma_5)+|B_3|^2(\gamma_5-\gamma_6)]$$

$$G_5 = \frac{1}{2Er_5}[|B_6|^2(\gamma_3-\gamma_4)+|B_3|^2(\gamma_4-\gamma_6)+|B_4|^2(\gamma_6-\gamma_3)]$$

$$G_6 = \frac{-1}{2E}[|B_3|^2(\gamma_4-\gamma_5)+|B_4|^2(\gamma_5-\gamma_3)+|B_5|^2(\gamma_3-\gamma_4)]$$

$$H_3 = \frac{-1}{Er_3}[\gamma_4(\delta_5-\delta_6)+\gamma_5(\delta_6-\delta_4)+\gamma_6(\delta_4-\delta_5)]$$

$$H_4 = \frac{1}{Er_4}[\gamma_5(\delta_6-\delta_3)+\gamma_6(\delta_3-\delta_5)+\gamma_3(\delta_5-\delta_6)]$$

$$H_5 = \frac{-1}{Er_5}[\gamma_6(\delta_3-\delta_4)+\gamma_3(\delta_4-\delta_6)+\gamma_4(\delta_6-\delta_3)]$$

$$E = \gamma_3(\delta_4-\delta_5)+\gamma_4(\delta_5-\delta_3)+\gamma_5(\delta_3-\delta_4) \tag{7.19}$$

7.2.2 Geometrical interpretation and design considerations

The equations describing six-port reflectometer operation have a geometrical interpretation which is helpful in designing six-port junctions, and in understanding the relationships between the six-port constants. We shall concentrate on examining the significance of the values B_i in eqns. 7.14, but any conclusions are equally applicable to the A_i in eqns. 7.7, since

$$A_i = \frac{-1}{B_i} \tag{7.20}$$

By using

$$|\Gamma-B_i|^2 = (\Gamma-B_i)(\Gamma-B_i)^* = (\Gamma-B_i)(\Gamma^*-B_i^*) \tag{7.21}$$

the ith equation in eqns. 7.14 can be written as

$$\Gamma\Gamma^*(e_i-1)+\Gamma(B_i^*-e_iB_6^*)+\Gamma^*(B_i-e_iB_6) + (e_i|B_6|^2-|B_i|^2) = 0 \tag{7.22}$$

where

$$e_i = \frac{p_i}{r_i} \tag{7.23}$$

The expression 7.22 describes a circle on the (complex) Γ plane, with radius R_i given by

$$R_i = (e_i)^{1/2} \left| \frac{B_6 - B_i}{e_i - 1} \right| \tag{7.24}$$

and centre C_i given by

$$C_i = \frac{e_i B_6 - B_i}{e_i - 1} \tag{7.25}$$

C_i, which lies on the straight line

$$\Gamma(B_6^* - B_i^*) - \Gamma^*(B_6 - B_i) + B_6 B_i^* - B_i B_6^* = 0 \tag{7.26}$$

passing through B_i and B_6, is beyond B_i if $0 < e_i < 1$, but beyond B_6 if $e_i > 1$. Members of the family of circles are shown in Fig. 7.7 for various values of e_i. If $e_i = 1$, the circle degenerates into a straight line which bisects normally the line between B_i and B_6.

When a termination with reflection coefficient Γ_L is connected to the reflectometer, each of the three equations in eqns. 7.14 defines a circle on the Γ plane, in accordance with eqns. 7.24 and 7.25. These three circles have (theoretically) a common point of intersection Γ_L, as shown in Fig. 7.8. In practice, because of imperfections in the calibration, and because of measurement noise, the three circles will not have a common point of intersection, but will define a quasi-triangular area on the Γ plane. The locations of the B_i on the plane are therefore important in keeping the area of this quasi-triangle, and consequently the uncertainty on the value obtained for Γ_L, small.

It is essential that the ratios p_i in eqns. 7.14 remain finite for all values of Γ which may be presented at the measuring port. For passive terminations with $|\Gamma| \leqslant 1$, the junction must therefore be designed, and port 6 chosen, so that

$$|B_6| > 1 \tag{7.27}$$

Geometrically, this means that B_6 must lie outside the unit circle on the complex Γ plane.

Another restriction on the B_i is evident from eqns. 7.19. If the magnitudes of all the B_i are equal, the expressions for the F_i and G_i collapse to zero. It is therefore essential that, geometrically, the B_i do not all lie on a circle with centre $\Gamma = 0$ on the complex Γ plane.

In practice, it is desirable that the dynamic range required of the power meters be not too large. If any $|B_i| \leqslant 1$, there are values of Γ which make the numerator in eqns. 7.14, and therefore p_i, small, and the particular value $\Gamma = B_i$ sets $p_i = 0$. The consequent demands on the dynamic range of the power meter are avoided if the desirable, but non-essential, restriction

$$|B_i| > 1, \quad i = 3, 4, 5 \tag{7.28}$$

is enforced. Geometrically, this requires that the B_i all lie outside the unit circle on the complex Γ plane. This restriction also has advantages in reducing the sensitivity of the result to noise. It can be seen from eqn. 7.24 that when p_i is small, the radius R_i of one circle is proportional to $(p_i)^{1/2}$. Any noise added to the power detected at port i causes a more significant error in R_i, and therefore Γ, than when p_i is large.

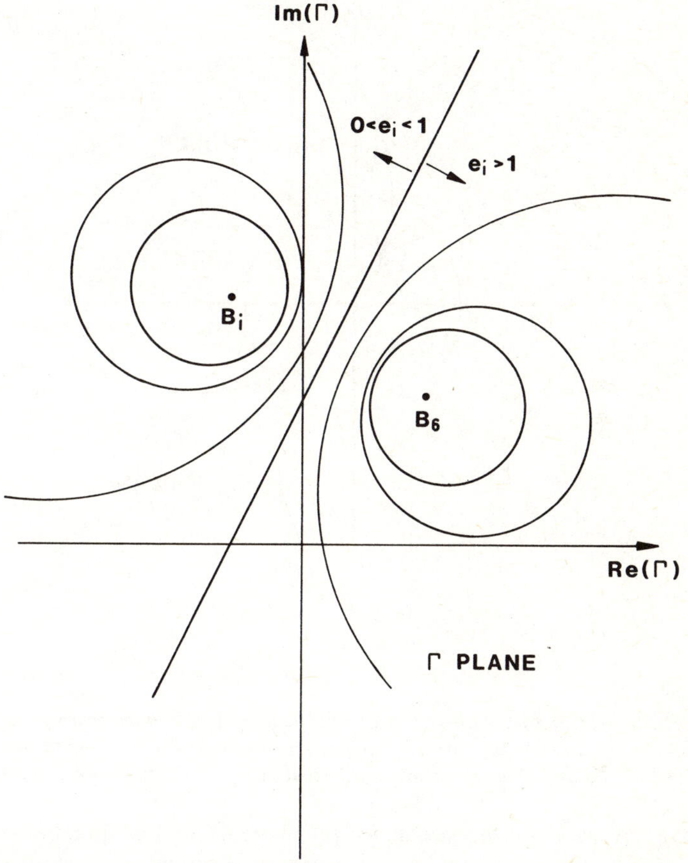

Fig. 7.7. *Different loci of Γ with different normalized ratios e_i of the power measured at one port of a six-port reflectometer to that measured at port 6. For a particular value of e_i, the locus of Γ on the complex Γ plane is a circle with centre on the line joining the B_i and B_6 of eqns. 7.14*

In particular, the desirable restrictions, eqns. 7.28, avoids the case $B_i = 0$, which has been argued against in detail by Engen [12] on the grounds of noise sensitivity when measuring a termination near a match, which is likely to be one of the most important uses of the reflectometer. Ensuring that no power meter responds solely to the reflected signal also allows the use of the equation form 7.7.

If, in spite of the above disadvantages, a six-port reflectometer is built with one $B_i = 0$, and equations of the form 7.14 are used to describe its behaviour, care must be taken to avoid the condition $M_i = 0$. This may be difficult at all frequencies, particularly if a directional coupler is used to monitor the generator output ($S_{i2} \simeq 0$), and the measuring port is well matched ($S_{22} \simeq 0$).

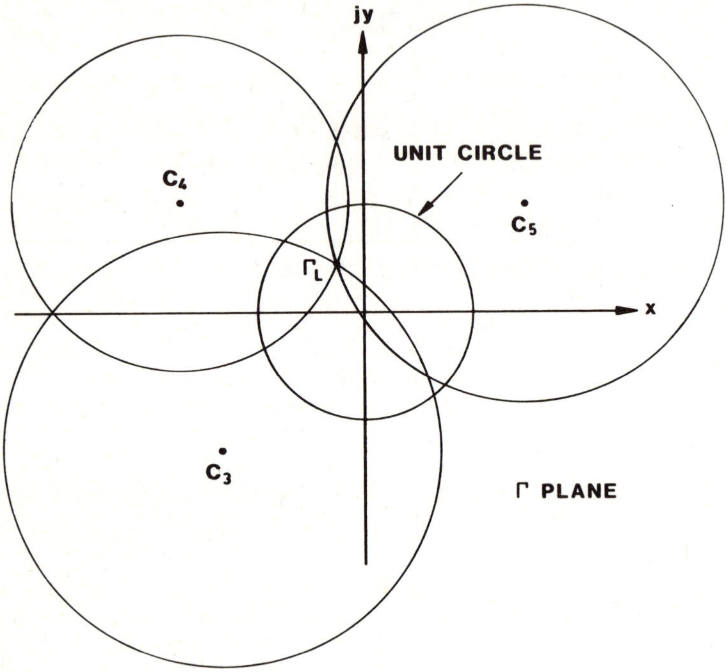

Fig. 7.8. *When a termination with reflection coefficient Γ_L is connected to port 2 of a six-port reflectometer, the ratios of the powers measured at ports 3–5 to that measured at port 6 define three circles which intersect at Γ_L on the complex Γ plane*

7.2.2.1 Six-ports with the forward signal monitored: Six-port junctions are often designed incorporating a directional coupler for monitoring the forward signal. Choosing the coupled port to be port 6 ensures that $|S_{61}| \gg |S_{62}|$, and therefore $|B_6| \gg 1$, provided that the measuring port mismatch is not excessive. The restriction 7.27 is therefore satisfied, and we can make the approximation

$$\frac{p_i}{r_i} = e_i \simeq \left|\frac{\Gamma - B_i}{B_6}\right|^2 \ll 1, \quad |B_6| \gg 1, |B_6| \gg |B_i| \qquad (7.29)$$

which leads to

$$\begin{aligned} R_i &\simeq |\Gamma - B_i| \\ C_i &\simeq B_i \end{aligned} \qquad (7.30)$$

The B_i, $i = 3, 4, 5$, therefore approximate the centres of the three circles on the complex Γ plane for all values of reflection coefficient presented at the measuring port.

It is arguable from symmetry considerations that the best locations for these circle centres are points equidistant from the origin of the complex Γ plane, but separated by 120°. It remains to select the magnitudes of the B_i.

The same arguments that led to eqns. 7.28 can be used to deduce that $|B_i|$ should not be too near unity, as then a p_i could be small for a fully reflecting termination. However, if the $|B_i|$ are large, it can be seen from eqn. 7.29 that a small change in p_i represents a large change in Γ. Geometrically, the radii R_i only need a small relative change to sweep from one side of the unit circle to the other, and cover the full range of Γ. Therefore choosing $|B_i|, i = 3, 4, 5$, to be large places high resolving demands on the power meters. A compromise value suggested by Engen [12] is for $|B_i|, i = 3, 4, 5$, to be in the vicinity of 1·5.

With $|B_6| \gg 1$ and $|B_i| > 1, i = 3, 4, 5$, the dynamic range required of the power meters is approximately

$$20 \log_{10} \left[\frac{|B_i| + 1}{|B_i| - 1} \right] \text{ dB} \tag{7.31}$$

With $|B_i| = 1{\cdot}5$, the dynamic range required is approximately 14 dB.

A detailed discussion of the merits and disadvantages of varous six-port junction configurations is beyond the scope of this book. The interested reader is referred to some of the many suggested configurations which have appeared in the literature [19, 25–33].

7.2.3 Calibration of six-port reflectometers

The purpose of calibrating a six-port reflectometer is to determine the constants F_i, G_i, H_i, either directly from eqn. 7.12 [19], or indirectly using a formulation such as eqns. 7.7 or eqns. 7.14. A succession of known terminations is attached to port 2, and the ratios of the power meter readings noted. Various calibration procedures have been suggested in the literature, differing mainly in the number and type of known standard terminations required [34–39]. These procedures include the use of four reactive standards [34], five general standards of which one is dissipative [35], four reactive standards with a matched load [36], four or five reactive standards with a matched load and an intermediate standard of known magnitude [19], and seven complex standards [25].

Each procedure has advantages and drawbacks. For example, the calibration using four offset short circuits [34] assumes that in each case $|\Gamma| = 1$. In practice there is loss associated with a short-circuit element and with its offsetting line, but the procedure does not allow for the incorporation of this information. In our experience of six-port precision reflectometry, the use of only reactive calibrating standards leads to reduced confidence in measurements taken near $\Gamma = 0$, whereas the use of a nominally matched load (precision sliding load, which in coaxial line should be beadless) as a calibrating standard tends to better define the behaviour of the six-port in that vicinity [40].

The following is a calibration procedure designed to determine the q_i and A_i of eqns. 7.7, from which the F_i, G_i, H_i, may be computed using eqns. 7.13 [35]. The calibration requires five standard terminations Z_k, $k = 1, \ldots, 5$, having reflection coefficients Γ_k in the desired Z_0 system of measurement. The impedances of the standards must neither all have the same magnitude, nor all have the same argument. The advantage of this calibration method is that it is explicit (non-iterative), and therefore rapid, yet still admits accurate descriptions of non-ideal standards.

The standards are realized conveniently (but not necessarily) as four practical (non-ideal) offset short circuits, and a load Z_5 which is near a match in a Z_0 system.

Step 1
Designate as standard number 5 the termination with reflection coefficient magnitude nearest to zero in the desired Z_0 system of measurement. Then for all five standards, compute

$$\Gamma'_k = x'_k + \mathrm{j}y'_k = \frac{\Gamma_k - \Gamma_5}{1 - \Gamma_k\Gamma_5}, \quad k = 1, 2, 3, 4, 5 \tag{7.32}$$

Note that $\Gamma'_5 \equiv 0$. If the four values $|\Gamma'_k|$, $k = 1, 2, 3, 4$, are similar, ensure that the two standards with the most positive values of x'_k are assigned the indices $k = 1$ and $k = 3$.

Step 2
Compute the four constants η_k:

$$\begin{aligned}
\eta_1 &= S_{1,2}C_{2,3}C_{3,4} - S_{2,3}C_{3,4}C_{4,1} - C_{1,2}C_{2,3}S_{3,4} + C_{2,3}C_{3,4}S_{4,1}\\
\eta_2 &= S_{2,3}C_{3,4}C_{4,1} - S_{3,4}C_{4,1}C_{1,2} - C_{2,3}C_{3,4}S_{4,1} + C_{3,4}C_{4,1}S_{1,2}\\
\eta_3 &= S_{3,4}C_{4,1}C_{1,2} - S_{4,1}C_{1,2}C_{2,3} - C_{3,4}C_{4,1}S_{1,2} + C_{4,1}C_{1,2}S_{2,3}\\
\eta_4 &= S_{4,1}C_{1,2}C_{2,3} - S_{1,2}C_{2,3}C_{3,4} - C_{4,1}C_{1,2}S_{2,3} + C_{1,2}C_{2,3}S_{3,4}
\end{aligned} \tag{7.33}$$

where

$$C_{n,m} = c_n - c_m; \quad S_{n,m} = s_n - s_m \tag{7.34}$$

and

$$c_n = \frac{x'_n}{|\Gamma'_n|^2}, \qquad s_n = \frac{y'_n}{|\Gamma'_n|^2} \tag{7.35}$$

Step 3
Connect successively to port 2 of the reflectometer each of the five calibrating standards. For each, record the ratio $p_{i,k}$ of the power measured at the ith port, $i = 3, 4, 5$, to that measured at the reference port, port 6.

Step 4
For each i, $i = 3, 4, 5$, compute

$$T_{i,k} = \frac{p_{i,k}}{p_{i,5}}, \quad k = 1, 2, 3, 4 \tag{7.36}$$

and then

$$e_i = \sum_{k=1}^{4} \frac{(T_{i,k} - 1)\eta_k}{|\Gamma_k'|^2}$$

$$f_i = \sum_{k=1}^{4} T_{i,k}\eta_k$$

$$g_i = 2\sum_{k=1}^{4} T_{i,k}c_k\eta_k$$

$$h_i = 2\sum_{k=1}^{4} T_{i,k}s_k\eta_k \tag{7.37}$$

Step 5
For each of the three combinations $i, j (i \neq j)$ obtainable from the numbers 3, 4, 5, compute

$$M_{i,j} = \frac{\xi_1^2 - 2(\xi_2\xi_3 + \xi_4\xi_5)}{2(\xi_2^2 + \xi_4^2)}$$

$$N_{i,j} = \frac{\xi_3^2 + \xi_5^2}{\xi_2^2 + \xi_4^2} \tag{7.38}$$

where

$$\xi_1 = g_i h_j - h_i g_j$$

$$\xi_2 = h_i f_j - f_i h_j$$

$$\xi_3 = h_i e_j - e_i h_j \tag{7.39}$$

$$\xi_4 = g_i f_j - f_i g_j$$

$$\xi_5 = g_i e_j - e_i g_j$$

Step 6
Compute

$$\alpha_6 = \frac{[M_{i,j} - (M_{i,j}^2 - N_{i,j})^{1/2}]\,\xi_2 + \xi_3}{\xi_1}$$

$$\beta_6 = \frac{[M_{i,j} - (M_{i,j}^2 - N_{i,j})^{1/2}]\,\xi_4 + \xi_5}{\xi_1} \tag{7.40}$$

The three values so obtained for α_6 and β_6 are redundant theoretically, but may be averaged in practice to reduce the effects of measurement noise. Then compute

$$|A_6|^2 = \alpha_6^2 + \beta_6^2 \tag{7.41}$$

Step 7

For each value of i, $i = 3, 4, 5$, compute four quantities $R_{i,k}$ given by

$$R_{i,k} = \frac{T_{i,k} - 1}{|\Gamma'_k|^2} + T_{i,k}(|A_6|^2 + 2\alpha_6 c_k - 2\beta_6 s_k), \quad k = 1, 2, 3, 4 \tag{7.42}$$

and then for each of the four combinations l, m, n $(l \neq m \neq n)$ obtainable from the numbers 1, 2, 3, 4, compute

$$\alpha_i = \frac{R_{i,l}S_{m,n} + R_{i,m}S_{n,l} + R_{i,n}S_{l,m}}{2(c_l S_{m,n} + c_m S_{n,l} + c_n S_{l,m})}$$
$$\beta_i = \frac{R_{i,l}C_{m,n} + R_{i,m}C_{n,l} + R_{i,n}C_{l,m}}{2(c_l S_{m,n} + c_m S_{n,l} + c_n S_{l,m})} \tag{7.43}$$

The four values so obtained for each α_i and β_i are redundant theoretically, but may be averaged in practice to reduce the effects of measurement noise.

Then compute

$$|A_i|^2 = \alpha_i^2 + \beta_i^2, \quad i = 3, 4, 5 \tag{7.44}$$

Step 8

Compute the F_i, G_i, H_i, using eqns. 7.13 with $q_i = p_{i,5}$.

This completes the calibration procedure. However, when an unknown termination is connected to the reflectometer, and eqn. 7.12 evaluated, the reflection coefficient Γ' obtained is that applicable in a Z_5 measurement system. The transformation to the desired Z_0 system of measurement is

$$\Gamma = \frac{\Gamma' + \Gamma_5}{1 + \Gamma'\Gamma_5} \tag{7.45}$$

Of course, if Z_5 is a matched load in the Z_0 system $(Z_5 = Z_0)$, then neither this final transformation nor Step 1 is necessary.

When the five calibrating standards are four offset (non-ideal) short circuits and a dissipative load, the short circuits may be realized as four positions of a sliding short-circuit element in a length of precision transmission line. Not only does a sliding short circuit require only one connection to be made to the measuring port of the six-port junction in order to present four calibrating standards, but the offset of each short-circuit position is readily adjusted to suit the frequency of calibration. The alternative, of using fixed lengths of precision transmission line terminated in short circuits, requires the separate connection of each standard to the measuring port, and the availability of many different lengths if the offset short-circuit positions are to be well chosen for each frequency of calibration.

However, knowledge of the characteristic impedance and propagation constant of the line in which a sliding shorting element moves is insufficient to define the electrical location of the short-circuit plane, since this is a function of the design of the shorting element, and the frequency. This difficulty can be overcome in the following manner.

The micrometer scale of the sliding short circuit is used to describe the short-circuit position thereby placing the phase reference plane at the electrically unknown position corresponding to the scale zero. Since, by definition, $\Gamma_5' = 0$, this choice of a phase reference plane does not affect the fifth calibrating standard. After the calibration has been completed, a high quality shorting element is connected to the measurement port, and its reflection coefficient Γ_s' measured. The magnitude of Γ_s' is a measure of the fixed contact loss of the sliding short-circuit element and the loss in the transmission line from the measurement port to the first sliding short-circuit position. The argument of Γ_s' is a measure of the displacement of the reference phase plane from the measurement port. The calibration of the six-port with respect to the measuring port is therefore calculable by rerunning the calibration algorithm using $-\Gamma_i'/\Gamma_s'$ in place of the original Γ_i'. After this, the six-port will measure the reflection coefficient of any unknown termination referenced to the plane of the fixed short circuit.

7.2.3.1 Alternative calibration: To calibrate a reflectometer in terms of the formulation 7.14 using the same standards as in Section 7.2.3, the procedure is identical for Steps 1, 2, 3.

Step 4

For each i, $i = 3, 4, 5$, compute

$$\begin{aligned}
e_i &= \sum_{k=1}^{4} p_{i,k}\eta_k \\
f_i &= \sum_{k=1}^{4} \frac{(p_{i,k} - p_{i,5})\eta_k}{|\Gamma_k'|^2} \\
g_i &= -2\sum_{k=1}^{4} p_{i,k}c_k\eta_k \\
h_i &= 2\sum_{k=1}^{4} p_{i,k}s_k\eta_k
\end{aligned} \tag{7.46}$$

Step 5 is as previously.

Step 6

Compute

$$\begin{aligned}
\gamma_6 &= \frac{[M_{i,j} + (M_{i,j}^2 - N_{i,j})^{1/2}]\,\xi_2 + \xi_3}{\xi_1} \\
\delta_6 &= \frac{[M_{i,j} + (M_{i,j}^2 - N_{i,j})^{1/2}]\,\xi_4 + \xi_5}{\xi_1}
\end{aligned} \tag{7.47}$$

The three values so obtained for γ_6 and δ_6 are redundant, but may be averaged in practice to reduce the effects of measurement noise. Then compute

$$|B_6|^2 = \gamma_6^2 + \delta_6^2 \tag{7.48}$$

Step 7

For each value of i, $i = 3, 4, 5$, compute four quantities $R_{i,k}$ given by

$$R_{i,k} = \frac{(p_{i,k} - p_{i,5})|B_6|^2}{|\Gamma_k'|^2} - p_{i,k}(2\gamma_6 c_k + 2\delta_6 s_k - 1), \quad k = 1, 2, 3, 4 \tag{7.49}$$

and then for each of the four combinations $l, m, n, (l \neq m \neq n)$ obtainable from the numbers 1, 2, 3, 4, compute

$$\tau_i = R_{i,l}(c_m s_n - c_n s_m) + R_{i,m}(c_n s_l - c_l s_n) + R_{i,n}(c_l s_m - c_m s_l) \tag{7.50}$$

and hence

$$r_i = \tau_i / (c_l S_{m,n} + c_m S_{n,l} + c_n S_{l,m})$$

$$\gamma_i = -(R_{i,l} S_{m,n} + R_{i,m} S_{n,l} + R_{i,n} S_{l,m})/(2\tau_i)$$

$$\delta_i = (R_{i,l} C_{m,n} + R_{i,m} C_{n,l} + R_{i,n} C_{l,m})/(2\tau_i) \tag{7.51}$$

The four values so obtained for each r_i, γ_i, δ_i, are redundant theoretically, but may be averaged to reduce the effects of measurements noise. Then compute

$$|B_i|^2 = \gamma_i^2 + \delta_i^2, \quad i = 3, 4, 5 \tag{7.52}$$

Step 8

Compute the F_i, G_i, H_i, using eqns. 7.19.

As before, the reflection coefficient calculated for an unknown termination using eqn. 7.12 is that applicable to a Z_5 system of measurement, and the transformation 7.45 is required to find the reflection coefficient in the desired Z_0 system of measurement.

References

1 MENDEL, E.: 'The system approach to scalar automatic network analysers', *Microwave J.*, 1983, **26**, pp. 49–70

2 BEATTY, R. W.: 'Automatic measurement of network parameters – a survey', *NBS Monograph*, 1976, no. 151

3 FITZPATRICK, J.: 'Error models for systems measurements', *Microwave J.*, 1978, **21**, pp. 63–66

4 BIANCO, B., CORANA, A., RIDELLA, S., and SIMCICH, C.: 'Evaluation of errors in calibration procedures for measurement of reflection coefficient', *IEEE Trans.*, 1978, **IM-27**, pp. 345–348

5 SALEH, A. A. M.: 'Explicit formulas for error correction in microwave measuring sets with switching-dependent port mismatches', *IEEE Trans.*, 1979, **IM-28**, pp. 67–71

6 WOODS, D.: 'Simplified calibration equations for the unidirectional and bidirectional computer-corrected network analyzer', *Proc. IEE*, 1980, **127**, pp. 82–86

7 FITZPATRICK, J. K.: 'Automatic network analyzer accuracy: How to get it, lose it, then regain it', *Microwave Syst. News*, 1980, **10**, pp. 70–93

8 WEINSCHEL, B. O. and WEINERT, F. K.: 'IF vector substitution increases accuracy', *Microwave Syst. News*, 1980, **10**, pp. 94–106

9 WARNER' F. L.: 'Microwave Attenuation Measurement', (Peter Peregrinus, 1977)

10 ENGEN, G. F. and HOER, C. A.: 'Application of an arbitrary 6 port junction to power-measurement problems', *IEEE Trans.*, 1972, **IM-21**, pp. 470–474

11 HOER, C. A.: 'The six-port coupler: a new approach to measuring voltage, current, power, impedance and phase', *IEEE Trans.*, 1972, **IM-21**, pp. 466–470

12 ENGEN, G. F.: 'The six-port reflectometer: an alternative network analyzer', *IEEE Trans.*, 1972, **IM-21**, pp. 1075–1080

13 HOER, C. A.: 'Using six-port and eight-port junctions to measure active and passive circuit parameters', NBS Tech. Note 673, 1975

14 ENGEN, G. F.: 'The six-port measurement technique – a status report', *Microwave J.*, 1978, **21**, pp. 18–89

15 ENGEN, G. F.; 'Thru-reflect-line': an improved technique for calibrating the dual six-port automatic network analyzer', *IEEE Trans.*, 1979, **MTT-27**, pp. 987–993

16 WOODS, D.: 'Analysis of a dual-six-port network analyzer and derivation of calibration equations', *Proc. IEE.*, 1980, **127**, pp. 541–548

17 CRONSON, H. M. and SUSMAN, L.: 'A dual six-port automatic network analyzer', *IEEE Trans.*, 1981, **MTT-29**, pp. 372–378

18 HOER, C. A.: 'A high-power dual six-port automatic network analyzer used in determining biological effects on RF and microwave radiation', *IEEE Trans.*, 1981, **MTT-29**, pp. 1356–1363

19 SOMLO, P. I. and HUNTER, J. D.: 'A six-port reflectometer and its complete characterization by convenient calibration procedures', *IEEE Trans.*, 1982, **MTT-30**, pp. 186–192

20 BRAUN, A. E.: 'Six-port instrumentation faces controversy', *Microwave Syst. News*, 1979, **9**, pp. 79–81

21 SIMMONS, J.: 'Six-port redux', *Microwave Syst. News*, 1980, **10**, p. 16

22 WEINSCHEL, B. O.: 'Six-port sceptic', *Microwaves*, 1980, **19**, p. 11–12

23 FONG-TOM, R. and CRONSON, H. M.: 'Diode detector characteristics for a 94 GHz 6-port application', *IEEE Trans.*, 1983, **MTT-31**, pp. 158–164

24 SOMLO, P. I., HUNTER, J. D. and ARTHUR, D. C.: 'Accurate six-port operation with uncalibrated nonlinear diodes', *IEEE Trans.*, 1985, **MTT-33**, pp. 281–282

25 WOODS, D., 'Analysis and calibration theory of the general 6-port reflectometer employing four amplitude detectors', *Proc. IEE*, 1979, **126**, pp. 221–227

26 ENGEN, G. F.: 'Design considerations for automatic network analyzers based on the six-port concept', Europ. Conf. on Prec. Electr. Meas., EUROMES-77, 1977, pp. 110–112

27 ENGEN, G. F.: 'An improved circuit for implementing the six-port technique of microwave measurements', *IEEE Trans.*, 1977, **MTT-25**, pp. 110–112

28 GROLL, H. P. and KOHL, W.: 'Six-port consisting of two directional couplers and two voltage probes for impedance measurement in the millimeter-wave range', *IEEE Trans.*, 1980, **IM-29**, pp. 386–390

29 RAFAL, M. D. and JOINES, W. T.: 'Optimizing the design of the six-port junction', *IEEE Trans.*, 1980, **IM-29**, pp. 437–439

30 PROBERT, P. J. and CARROLL, J. E.: 'Design features of multiport reflectometers', *Proc. IEE*, 1982, **129**, pp. 245–252

31 HANSSON, E. R. B. and RIBLET, G. P.: 'An ideal six-port network consisting of a matched reciprocal lossless five-port and a perfect directional coupler', *IEEE Trans.*, 1983, **MTT-31**, pp. 284–288

32 TARR, L. W.: 'An automated six-port for 2–18 GHz power and complex reflection coefficient measurements', *IEEE Trans.*, 1983, **IM-32**, pp. 162–164

33 GRIFFIN, E. J., SLACK, G. J. and HILL, L. D.: 'Broadband six-port reflectometer junction', *Electron. Lett.*, 1983, **19**, pp. 921–922

34 LI, S. and BOSISIO, R. G.: 'Calibration of multiport reflectometers by means of four open/short circuits', *IEEE Trans.*, 1982, **MTT-30**, pp. 1085–1090

35 HUNTER, J. D., and SOMLO, P. I.: 'An explicit six-port calibration method using five standards', *IEEE Trans.*, 1985, **MTT-33**, pp. 69–72

36 RIBLET, G. P. and HANSSON, E. R. B.: 'Aspects of the calibration of a single six-port using a load and offset reflection standards', *IEEE Trans.*, 1982, **MTT-30**, pp. 2120–2125

37 ENGEN, G. F.: 'Calibrating the six-port reflectometer by means of sliding terminations', *IEEE Trans.*, 1978, **IM-26**, pp. 951–957

38 ENGEN, G. F.: 'Calibration of an arbitrary six-port junction for measurement of active and passive circuit parameters', *IEEE Trans.*, 1973, **IM-22**, pp. 295–299

39 EL-DEEB, N. A.: 'The calibration and performance of a microstrip six-port reflectometer', *IEEE Trans.*, 1983, **MTT-31**, pp. 509–514

40 SOMLO, P. I.: 'The case for using a matched load standard for six-port calibration', *Electron. Lett.*, 1983, **19**, pp. 979–980

Appendixes

Appendix A

Numerical flow graph reduction

The following is an implementation in the BASIC language of the reduction of flow graphs, as described in Section 2.3.3.

After specifying the number of nodes, N, of the flow graph, two matrices A and B of order $N+1$ are set up. Element $A(i, j)$ is the real part of the branch value between nodes i and j, and element $B(i, j)$ is the imaginary part. After nominating the input and output nodes I1 and O1, respectively, column $N+1$ and row $N+1$ of both A and B are set to zero, with the exceptions

$$A(N+1, \mathrm{I1}) = 1 \quad \text{and} \quad A(\mathrm{O1}, N+1) = 1.$$

The flow graph matrix reduction is then carried out in the following manner:

```
100 K=0
110 K=K+1
120 FOR I=K+1 TO N+1
130 IF (A(I,K)=0 AND B(I,K)=0) THEN GOTO 240
140 FOR J=K+1 TO N+1
150 IF (A(K,J)=0 AND B(K,J)=0) THEN GOTO 230
160 R7=A(I,K)*A(K,J)-B(I,K)*B(K,J)
170 I7=A(I,K)*B(K,J)+B(I,K)*A(K,J)
180 R8=1-A(K,K)
190 I8=-B(K,K)
200 D1=R8*R8+I8*I8
210 A(I,J)=A(I,J)+(R7*R8+I7*I8)/D1
220 B(I,J)=B(I,J)+(I7*R8-R7*I8)/D1
230 NEXT J
240 NEXT I
250 IF K<N THEN GOTO 110
260 exit. . .
```

After exiting at 260, the real and imaginary parts of the transmission from node I1 to O1 are available in $A(N+1, N+1)$ and $B(N+1, N+1)$, respectively. If nodes I1 and O1 are both associated with the same network port, the result of the computation is the reflection coefficient seen at that port.

The checks for zero elements in statements 130, and 150 are to save computational time, since in most cases the matrices are rather sparse, because not every node is connected to every other node by a branch.

If the original matrices are saved, new branches may be added, and changes made to existing branches and input and output nodes, before computing another transmission coefficient. New nodes may not be created without increasing the size of the matrices.

In practice, it is convenient to organize the program to assemble the matrices from predefined 'building blocks' such as the flow graph representations of series and shunt immittances, and sections of transmission line, including waveguides. If these are defined as functions of frequency, it becomes a simple matter to program for stepped frequency analyses.

Appendix B

Measurement uncertainty

Practical microwave measurements are corrupted by errors which have their origin in random processes, and by errors caused by systematic bias. This appendix outlines how such errors can be accommodated when presenting measurement results. The terms used here are as defined in references 1 and 2.

(*a*) Uncertainty owing to random fluctuations

In a situation where the observed quantity x is corrupted by random fluctuations, repeated observations will produce different results. Before analysing the significance of these differences, it is necessary to know the probability distribution of the observations. Since it is impossible to take the infinite number of observations needed to find the true probability distribution, a likely probability distribution must be assumed. For most microwave measurements it can be argued that the probability distribution is approximately Gaussian or 'normal', with a probability distribution function

$$f(x) = \frac{1}{(2\pi)^{1/2}\sigma} \exp\left[-\frac{(x-\mu)^2}{2\sigma^2}\right] \tag{B.1}$$

where $f(x)\delta x$ is the probability of an observation lying within the range δx centred on $x - \mu$. μ, the population mean, is the mean of an infinite number of observations, and σ is the population standard deviation. Fig. B1 shows $f(x)$ for various values of σ. Any one observation must lie between $-\infty$ and $+\infty$, so the area under each of these curves is unity. Although, in practice, repeated measurements of a microwave quantity will never produce the extreme values possible with a Gaussian distribution, such a distribution is often a useful approximation to the true probability distribution of practical measurements.

σ, the population standard deviation, is defined for an infinite number of observations. For the finite sample of n observations x_i of the population, the unbiased estimate of σ is s, given by

$$s = \left[\frac{\sum_{i=1}^{n} (x_i - \bar{x})^2}{n-1} \right]^{1/2} \tag{B.2}$$

where $\bar{x}$ is the sample mean.

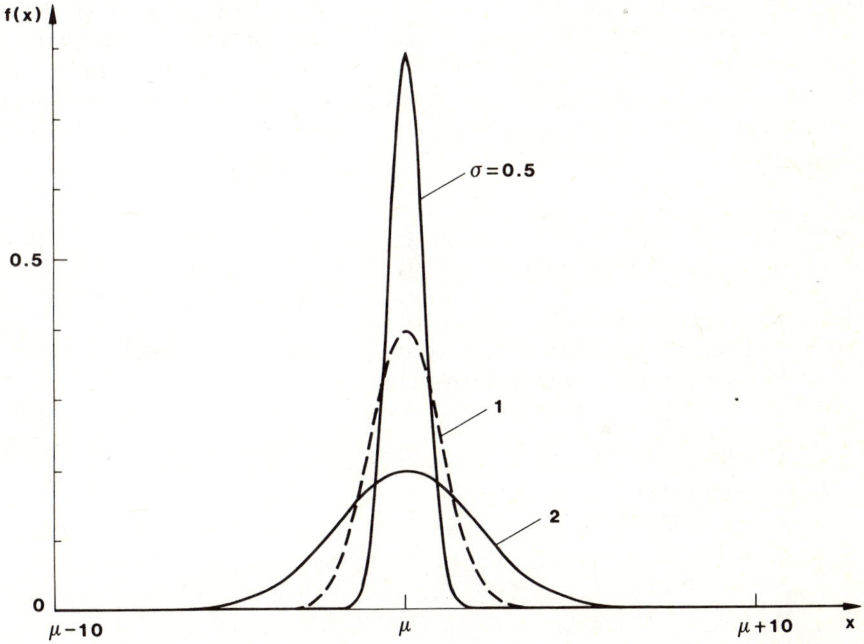

Fig. B.1. *Gaussian probability distributions with mean μ, for various values of the standard deviation σ*

If the sample mean and estimate of the standard deviation are re-evaluated as the number of observations is increased, the fluctuations in their values will diminish (provided the measured parameter is 'well-behaved'). A metrologist is interested in estimating how closely the sample mean $\bar{x}$ approaches the true (population) mean μ. This estimate can be described in terms standard deviation of the mean, s_m, which is a measure of the fluctuations of the sample mean after n observations, and is given by

$$s_m = \frac{s}{n^{1/2}} = \left[\frac{\sum_{i=1}^{n}(x_i - \bar{x})^2}{n(n-1)}\right]^{1/2} \tag{B.3}$$

The uncertainty U_r associated with the estimate $\bar{x}$ of μ depends not only on s_m, but also on the level of confidence with which the metrologist wishes to assert that μ lies within the range

$$\bar{x} - U_r \leqslant \mu \leqslant \bar{x} + U_r \tag{B.4}$$

The value of the uncertainty which takes into account these considerations is

$$U_r = t_{CL} s_m \tag{B.5}$$

where t_{CL} is the Student's t-factor for a given confidence level [1, 3], and is tabulated in many statistical texts as a function of the number of observations n.

For a simple computer calculation of uncertainty limits, it is convenient to have available an approximate formula for the Student's t-factor. Many metrologists quote a 99% confidence interval with the sample mean, and an empirical formula for $t_{99\%}$ is

$$t_{99\%} \simeq 2{\cdot}58 \left[\frac{n - 0{\cdot}3}{n - 2{\cdot}3}\right], \quad n \geqslant 3 \tag{B.6}$$

which is in error by less than 5%.

As an example of the calculations concerning measurements subject to random fluctuations, consider the following observations listed:

n	x
1	0·2321
2	0·2312
3	0·2315
4	0·2318
5	0·2325
6	0·2318
7	0·2320
8	0·2311

The sample mean is

$$\bar{x} = \frac{1}{8}\sum_{i=1}^{8} x_i = 0{\cdot}2318$$

with the unbiased estimate of σ, from eqn. B.2;

$$s = 0{\cdot}005$$

and s_m, from eqn. B.3

$$s_m = 0.0002$$

Using eqns. B.5 and B.6, the uncertainty (99% confidence level) associated with $\bar{x}$ is

$$U_r = 0.0006$$

The result may therefore be stated as

$$\mu = 0.2318 \pm 0.0006$$

at a confidence level of 99% which means that the metrologist is 99% confident that μ lies within the interval

$$0.2312 \leqslant \mu \leqslant 0.2324$$

(*b*) Uncertainty owing to systematic bias
In a situation where the measured quantity is subject solely to systematic bias, repeated measurements will produce the same (erroneous) result. The effect of the bias is to offset the measured result from the true value, and the bias cannot therefore be observed. (If an offset can be measured by some means, it can be removed from the result, and no longer contributes to the uncertainty). The effect of bias can only be estimated, after informed consideration of all possible contributing sources. These include bias caused by the measuring apparatus, the operating procedure, and the device under test.

It is often assumed that the components a_i of the systematic uncertainty, which arise from the various possible sources of bias, have rectangular probability distributions. These distributions are the result of the metrologist estimating the limiting values for each a_i, but being unable to argue that any particular value between these limits is more probable than any other value. The standard deviation for a series of rectangular distributions is

$$\sigma_s = \left[\frac{1}{3}\sum_{i=1}^{n} a_i^2\right]^{1/2} \tag{B.7}$$

(If triangular distributions were assumed, the denominator in eqn. B.7 would be 6 instead of 3).

Application of the Central Limit Theorem [3] suggests that it is reasonable to assume a Gaussian distribution as the overall probability distribution function for the effects of systematic bias. The uncertainty U_s can then be found by multiplying σ_s by the limiting ($n = \infty$) value of the Student's t-factor, which for a 99% confidence level is 2·576.

(*c*) Overall uncertainty
The overall uncertainty U_t is

$$U_t = [U_r^2 + U_s^2]^{1/2} \tag{B.8}$$

When the observations are of complex quantities, the statistical operations must be performed on the real and imaginary parts separately. The uncertainties associated with the real and imaginary parts of the results can be regarded as the components of a complex uncertainty, and the confidence interval will be an ellipse on the complex plane. If the two component uncertainties are equal, this ellipse will reduce to a circle.

References

1 'Guidelines for estimation and statement of overall uncertainty in measurement results', Commonwealth Science Council Publ., CSC(80)MS-9, 1980

2 'Glossary of metrological terms', Commonwealth Science Council Publ., CSC(80)MS-8, 1980

3 FREUND, J. E. and WALPOLE, R. E.: 'Mathematical Statistics' (Prentice-Hall, 1980)

Author index

Subject index